World University Library

The World University Library is an international series
of books, each of which has been specially commissioned.
The authors are leading scientists and scholars from all over
the world who, in an age of increasing specialization, see the
need for a broad, up-to-date presentation of their subject.
The aim is to provide authoritative introductory books for
students which will be of interest also to the general
reader. Publication of the series takes place in Britain,
France, Germany, Holland, Italy, Spain, Sweden and
the United States.

Frontispiece The system of the world according to the ideas prevalent in the Middle Ages. Above the flat Earth the sky forms a vault, through which the intrepid traveller pokes his head to discover the complicated mechanism moving the stars.

Jean Charon

Cosmology

Translated from the French
by Patrick Moore

World University Library

McGraw-Hill Book Company
New York Toronto

© Jean Charon 1970
Translation © George Weidenfeld and Nicolson Limited 1970
Library of Congress Catalog Card Number 73–77023

Photoset by BAS Printers Limited, Wallop, Hampshire, England
Manufactured by LIBREX, Italy

Contents

Introduction

It may be said that there are some kinds of questions which can never be given answers truly satisfactory to the human mind. We have limited brains, and we are busy with everyday matters, and so these questions are often neglected. Yet it is worth noting that scholars cannot provide rational answers to many questions; opinions and theories have varied tremendously during the course of human history, until by now science will not propose anything which is comprehensible merely to satisfy our sense of curiosity. Theological questions come into this category; it is impossible to answer them by sober formulae. Any attempt to evade the questions by dismissing them as false problems is tantamount to admitting that the questions have not been understood.

But the questions are there and are deeply rooted in our subconscious mind, waiting to be dealt with at some future time. Inevitably, these questions produce a sense of uneasiness; there is so much that we do not know.

What are we in this great universe? What should we be trying to do in order to make the best use of our brief lives? What is the status of the Earth in the universe? Is the universe finite? Did our world have a beginning – and if so, what came before? Who is the creator of all things? What is the course of evolution of the universe, and is the current of time sweeping us on toward annihilation or eternity? And when we die, does this mean our disappearance into absolute nothingness? Can we find any justification for the hope that the universe maintains its logic and its significance not only on a collective scale, but also on a personal one?

Man knows that questions of this kind are the basis of his existence. This is why we try so hard to find answers which will reconcile the known facts with what we feel about these problems. Admittedly, our replies are never definitive, and are always rather unsatisfactory, but they do at least lead to a slightly better understanding of things. Certainly, man feels the need to summarise everything that he has found out about the universe in space and in time, and he is doing his best.

Cosmology may be said to be the sum total of all the answers that we have found up to the present moment.

It is often claimed that in our scientific age our knowledge has been built up in a purely objective fashion; that is, by a description of nature and the way it works divorced from human personality with its irrational element. Yet this is not the case. Science is made not by science but by scientists. As soon as we approach problems which are at the very limits of our present knowledge, then scientists, with their own particular personalities, interfere with the observed phenomena and interpret them in a way which is sometimes emotional rather than strictly logical.

This drawback is particularly evident when we come to consider the universe as a whole. In this, more than anywhere else in science, man projects himself in the image that he believes he can see objectively in the world in which he lives. To show the truth of this, examine the different cosmologies which have been proposed one after the other. It is apparent that knowledge is built on metaphysics just as much as it is on physics.

It is undeniable that human knowledge increases steadily with time. All through history, knowledge of the universe has developed both by the collection of facts and by the quality and precision of the way in which these facts can be expressed. Much work was needed; there was steady improvement in the language of description, notably by means of mathematics.

During this steady progress in our knowledge of the universe as a whole, there were a few breakthroughs – sudden advances in human thought in which the whole problem of the universe was re-examined and all previous ideas modified. These changes were drastic. They affected the basis of thought, the methods of research, and the language of expression.

Two of these breakthroughs were particularly significant. The first came in the sixth century BC, when all human thought took on a new brilliance. This was the century of Buddha, Confucius and Lao-Tse; it was also the century of the 'Greek miracle', with the Ionian philosophers and Pythagoras.

Before the sixth century BC, cosmology simply enabled man to picture a comfortable, reassuring universe. Thus, like a child nestling in the arms of its mother, mankind could feel well protected against the immensity outside the Earth; there was no question of trying to make cosmology fit in with the observable facts. To the Babylonians, the Egyptians and the Hebrews, the world was an oyster floating in an immense sea; there was water above the sky (the upper shell) and water below the Earth (the lower shell). True, the shape of the oyster was discussed; it was round to the Babylonians but somewhat rectangular to the Egyptians, which, whatever may be the cause, is a curious shape for any oyster. These cosmological models were adorned with complex and often poetic details. Thus, the Egyptians thought that the sky must be the stomach of a cow which was standing on the rectangular world, its four feet at the four corners, or, alternatively, the belly of a woman who was supported on her elbows and her knees.

Above the ground, the stars danced in the manner of gods,

following a natural rhythm which was not the same as the rhythm of human life. Yet certain regularities were found. For instance, each month the Moon god was attacked by a fierce sow, which ate him steadily for two weeks; after this, the Moon god was reborn to a new existence. Then there were rhythms which were comfortable and soothing: day and night, seasons and rainfall, harvest and sowing time – all these governed life on Earth. The mythological outlook softened the hard contrasts of a world which man did not understand, a world which was both marvellous and frightening.

Then, quite abruptly, man discovered a new way of looking at the world. The transformation came in the sixth century BC, along the coast of the Aegean Sea, in Ionia. The new outlook was *reason*. Man found that the world had not been created specifically for him, and that explanations in accord with logical argument would allow him to understand at least some of the observed phenomena. When this had been attempted, man could sense the joy of being more completely united with nature. The new outlook made it possible to explore things which had previously been inaccessible; man no longer felt isolated and cut off. Even better, he was beginning to sense a tremendous ambition; he wanted to dominate this mysterious nature into which he had been born – an ambition that has remained ever since.

The rational outlook had been established. Over the years it was destined to develop, and to lead on to a new breakthrough at the beginning of our own century. During this long period, man used 'rational thought' to draw up numerous models, or cosmologies, of the universe which will be described in Part 1 of this book. All these models had one thing in common: they aimed to combine the logical deductions offered by reason with practical observations of the world and the universe beyond. Sometimes these cosmologies were based on principles which seemed eminently reasonable then, even if not now. As an example, take the axiom according to which all celestial bodies must move in circular orbits – because a circle is the perfect curve, and celestial bodies, which are divine, cannot move in anything but a perfect way. This particular axiom served as the basis for most cosmologies until the sixteenth century AD.

Newton altered matters. To him, observation came first, and all rules and logic had to be subordinated to the observed facts. But this did not alter the main theme, according to which reason and observation were both trying to make the deductions fit in with each other so as to provide a description of natural phenomena based on rational causes.

'Then came a change, as all human things change', to quote from Tennyson's poem 'Enoch Arden'. To the straightforward rational outlook came the relativistic outlook, whose originator was Albert Einstein. The perspective adopted for looking at our universe was suddenly altered in such a way that all the old cosmologies seemed to become totally inadequate. For the first time cosmology, the study of the universe as a whole, could be formulated in a manner as precise as that used to describe everyday phenomena on Earth. It became possible to speak scientifically of a universe which was both finite and unbounded; of a world having a beginning without ever having begun; of a world filled with immense energy without the need to conjure up a creator to account for it. Of course, the relativistic outlook did not involve the wholesale rejection of the rational outlook; indeed, very much to the contrary. After all, the rational outlook had not suddenly destroyed the mythological outlook which had preceded it. As we have noted, the myth of circular orbits for all celestial bodies had been used in a truly 'rational' manner according to the state of knowledge at the time. The relativistic outlook had merely built upon the foundations of the rational outlook, so that men could see better and further.

How exactly can this new form of relativistic thought be defined? This is the subject of part 2 of this book. For the moment, it will suffice to indicate the change in perspective to which it corresponds.

The relativistic outlook furnished an entirely new cosmology, because it brought with it a language capable of describing the view of the universe which would be had by a sort of 'super-observer', who was assumed to be outside the universe and therefore able to consider the whole of it, with its limitations of space and time, to say nothing of man set into the general framework. The change in perspective is rather the same as that of a man who climbs above the trees in a forest so that he can see the countryside and the roads beyond.

The rational outlook taught that things were either white *or* black, according to the viewpoint from which they were studied. Curiously enough, the relativistic outlook showed that things could be both white *and* black; it provided a unified, generalised language in which apparent contradictions were reconciled and were seen to be nothing more than complementary qualities of the same object.

During the past fifty years, the relativistic outlook has been of particular help in two principal ways. Rather in the manner of *Alice in Wonderland*, our super-observer can take an overall view of the universe, and it is this outlook which has led to the relativistic cosmologies described in this book. Yet the super-observer can also reduce his size so as to examine the almost infinitely small as well as the almost infinitely large. On the minute scale of the world of elementary particles, 'the modern scientist can no longer know whether what he observes is the essence of matter or the reflection of his own thought' (P. Teilhard de Chardin in *Le Phénomène humain*). The scientist must know the structure of thought, so that he can give the best possible interpretation of what he observes.

In orienting scientific language toward modern algebra, the relativistic outlook has also brought an entirely new interpretation of the world of the almost infinitely small. This aspect cannot be discussed here, but undoubtedly the interest and fascination of 'micrology' is as great as that of the study of cosmology.

Let us now go back twenty-five centuries, to the time when the first rational cosmologies were drawn up. As Aristotle so wisely said, 'To see things clearly, we must take them from their beginnings.'

Part 1

Rational Cosmologies

1 A world created by God

The question: 'Who created the world?' is apt to be embarrassing, even when put to an eminent modern scholar. The answer that 'the world was created by God' would almost unanimously have been given several centuries ago, and even today this reply has not been replaced by anything much clearer.

However, the development which took place during the sixth century BC was the formulation of the idea that God had created the world not in a purely haphazard manner, but in such a way that there were various different principles, resulting in the association of similar causes with the same effects. The idea of *principle* is essential in the development of human thought, because human logic is capable of functioning as soon as it has certain basic principles to work upon; and if the universe contains a certain number of fundamental principles, man can start to understand his surroundings even in spite of the apparent disorder. This is particularly the case if the principles are assumed to be valid over the entire universe, because knowledge of the observable phenomena in our own surroundings will be a safe guide to all the rest. The result was the building-up of a rational cosmology, together with a possible model for the universe.

Everything depended upon the validity of the principles which were thought to hold good for the whole universe. Changing the basic principles also involved changing cosmology, but without sacrificing what had come to be recognised as the rational outlook. Taking the cosmological problem into the realm of principle and reason meant, in fact, turning it into a science.

I believe that Pythagoras of Crotona should be regarded as the man who initiated rational cosmologies. His basic principles were followed by his successors right up to the time of Newton, and there was no attempt to break away from what may be called *reason*.

Pythagoras

Pythagoras was born in Greece some time during the first years of the sixth century BC. We have very little precise information about his life, though he seems to have lived to a great age, and to have been over ninety when he died. Early in his career he left his native Ionia and travelled; we know that he journeyed in Egypt and Babylonia. On his return to the island of Samos he fell foul of the Persians and went into voluntary exile with his family, settling in the Greek city of Crotona in Central Italy. It was there that he founded what we usually call the Pythagorean Brotherhood. By then he was very famous, and he had numerous disciples, but near the end of his life the leading politicians became alarmed by his spiritual influence, and he was banished from Crotona; his disciples were exiled, and their meeting place burned. All this added to the Pythagoras legend, and increased his spiritual influence over the following generations.

Curiously enough we have no writings of Pythagoras himself, probably because they were deliberately destroyed, but his fame was such that references to him are given by most of the pre-Socratic authors. Xenophanes mentions him once, Heraclitus three times, Empedocles once, and so on.

The fundamental principle of all Pythagorean philosophy is the idea of a correspondence between numbers and the mechanism of nature. To Pythagoras, whole numbers seemed to be capable of expressing the whole of nature's order and equilibrium. He seemed to be specially struck by the fact that if the strings of a lyre are put under equal tension, and then cut to lengths worked out in the form of quotients of whole numbers, they would produce the notes of the musical scale. This concept of 'rational numbers' became the basis of his description of the whole universe. It was not only understandable, but it created – as in music – the harmony of the world. Pythagoras might well have used the words which were actually written, some twenty-five centuries later, by the celebrated mathematician Léopold Kronecker (1823–91): 'God created natural numbers; the rest were created by Man.' In passing, it may be noted that Pythagoras chose a basic principle for his cosmology which was followed for a very long time indeed.

The idea of number, and the properties of numbers, should be considered together with the attempts to explain physical phenomena; there is a great deal of justification in the oft-expressed view that Pythagoras was the father of science. He gave the first clear indication of the fundamental method of science; he also outlined the representation of phenomena by means of mathematical formalism. This idea has prevailed; nowadays it is not unusual for a phenomenon to be given only a mathematical description, with no direct link with the data as given by our senses.

Pythagoras built up his cosmology solely by incorporating numbers in the way they were introduced in the harmony of a lyre. To him, the order of the universe seemed to prove that the suns and planets travelled along their orbits not in a haphazard manner, but in such a way that their movements created a celestial harmony – just as the lengths of the strings of a lyre created a musical harmony.

The first stage in Pythagorean cosmology was to admit that the Earth was a sphere, and not a disc floating in air (as Pythagoras had originally believed, in agreement with

Anaximander). Pythagoras then went on to suppose that all seven of the 'wandering' bodies – the Sun, the Moon, and the five planets then known: Mercury, Venus, Mars, Jupiter and Saturn – turned round the Earth in concentric circles, each of which was fixed to a sphere or a wheel. Beyond all these there came another sphere, that of the fixed stars. Air was thought to fill all the heavens; in disturbing the air, these various bodies produced musical notes, just as the strings of a lyre will do when lightly plucked. The concert due to these moving bodies was called the harmony of the spheres. It was as if the universe was made up of an immense lyre with circular strings; the laws of harmony, as known to the Pythagoreans, were also the laws of the whole universe. In his *Natural History*, Pliny says that according to the Pythagoreans, the musical interval formed between the Earth and the Moon was one tone; from Mercury to Venus, a semitone; from Venus to the Sun, a minor third; from the Sun to Mars, a tone; from Mars to Jupiter, a semitone; from Jupiter to Saturn, a semitone; and from Saturn to the sphere of the fixed stars, a minor third.

Certainly it was clear that this harmony of the spheres would be audible only to a very few people. It was claimed that nobody apart from the master, Pythagoras, could really hear the music made by these worlds in movement around our Earth.

Yet it would be wrong to laugh at Pythagorean cosmology; it was after all, the first to concede that the movements of celestial bodies obey certain laws, and that these rules can be expressed numerically just as with the lengths of the strings of a lyre. This alone represented real progress. The idea of definite order in natural phenomena is the first postulate of science.

Philolaus

By the middle of the fifth century BC, the Pythagorean community had run into difficulties. In particular, it had been realised that natural laws cannot be expressed simply by the relationships of whole numbers. For example, if the side of a square is given in any particular number of units, there will never be an exact number of the same units in the diagonal of the square. It is essential to bring in what is now called irrational numbers; the diagonal of a square is equal to the side multiplied by the square root of 2 ($\sqrt{2}$), but this root, which is itself a number, cannot be represented by the quotient of any two whole numbers. If so, it seemed that the same would hold good for the relationships between the radii of the orbits of celestial bodies – which at once disposed of the analogy with the notes of a musical scale. It also seemed to deal a death-blow to the idea of the harmony of the spheres.

Moreover, there were other reasons for the dissolution of the Pythagorean School. The brotherhood was preoccupied with the notion that the universal laws of harmony could be applied to moral and social problems on Earth, and they had freely put into practice a certain number of egalitarian principles, most notably the emancipation of women. This had alarmed the political authorities. The brotherhood was dissolved, and most of its members arrested. The leading disciples of the master (among whom the best known are Philolaus and Lysis) were, however, later allowed to return to Italy and continue with their teaching.

Philolaus was the first to question the theory that the Earth lies at the centre of the world system. He believed that the true centre must be somewhere else, and he pictured a

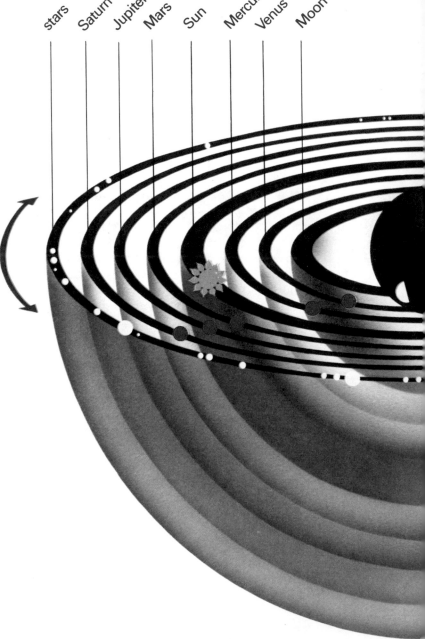

stars   Saturn   Jupiter   Mars   Sun   Mercury   Venus   Moon

1·1 The Pythagorean system, in which the Sun, Moon and the planets revolve around the central Earth. The fixed stars form the outermost 'shell'.

Earth

central fire, around which revolved first the Earth, then the Moon, followed by the Sun and the five planets; beyond came the sphere of the fixed stars, which did not revolve. As the Earth moved round the central fire, the stars in the motionless outer sphere would seem to move in the sky – which, of course, was exactly what could be observed. Beyond the outermost sphere was another immense fire; the stars were merely small holes in the outer sphere, so that the fire could be seen through each hole.

The most important part of Philolaus' cosmology was his belief that the Earth moved in space in the same way as other celestial bodies. The Earth retained a privileged position, since it lay closest to the central fire, but it was no longer the centre of all things.

Heraclides

The world system of Philolaus was based on the idea of the Earth's movement round a central fire; this certainly explained the daily movements of the fixed stars, since it was thought that the Earth itself was in motion rather than the outermost sphere. Yet in one way, the central fire was a little like Pythagoras' music of the spheres; it could not be seen, any more than the celestial harmony could be heard. It was Philolaus' successor. Heraclides of Pontus, who put forward a more reasonable theory, and suggested that the Earth, lying in the centre of the universe, rotated on its axis.

This, of course, was a major step forward. Also, the evidence indicated to Heraclides that at least two of the planets – Mercury and Venus – should be regarded as in motion round the Sun rather than round the Earth.

It was only one step more to propose that all the planets,

including the Earth, must move round the Sun. This idea duly came, chiefly from a Greek 'Copernican' named Aristarchus, but only about the year 275 BC – that is, almost two centuries after the time of Heraclides. During these two centuries a new way of considering the cosmological problem had taken root in Greece, due to the influence of Plato and Aristotle. This new picture, which put back the Earth in the centre of the world system, was followed by orthodox scientists for more than fifteen centuries, despite the intuition of Aristarchus. We must now examine the reasons why the Earth-centred view of the universe held sway for so long.

Plato

All through the history of science – particularly with ancient science – we find an alternation between the wish to build up knowledge by means of theory, and the wish to build it upon observation. Pythagoras had been one of the first to look for a principle which would guide all cosmological phenomena, and in his rational numbers, expressed as clearly for celestial orbits as for musical notes, he thought he had found it. After Pythagoras, Philolaus and Heraclides came back to sheer observation; to them, it was necessary to know how the various bodies revolved round one another, and what were their positions and their relative movements. The problem of why they moved, and what principles were involved, became of secondary importance.

With Plato, we return to science *a priori*. The first need is to look for an universal principle capable of explaining the world system; once this has been done, all ideas can be linked with this principle, and the principle itself can be used to provide complete explanations.

The basic idea is quite straightforward. The force which has put the celestial bodies in motion, and assigned to them the paths they follow, must be God, because it is God who has created the universe. But by definition, God is a perfect being. The movements he has conferred on the various bodies must therefore be perfect, and the orbits must be of perfect form.

Yet what is a perfect curve? Today, the question is very involved, but to the ancients there can be no doubt that the sphere was the perfect form, and the circle was the perfect curve. A circle has neither beginning nor end, so that it well symbolises the eternal association with God. And in any case, why should God choose to make the celestial bodies move in tortuous paths? How can we justify the idea that a body should be made to move sometimes to the right, sometimes to the left? The fundamental principle is quite clear cut: the celestial bodies can move only in circles. This is the dogma of the perfect circle, and it was regarded as being an axiom, just as modern physicists are forced to follow certain axioms which lie at the root of all natural phenomena. And if this axiom is taken as being valid all over the universe, it can be used to build up an axiomatic cosmology which is as true as the axiom itself.

With his laws of harmony, Pythagoras had already made use of the circle axiom. He decided in favour of a spherical Earth, rather than the disc-shaped world favoured by his predecessors, because he regarded the Earth as a divine creation, whose form could only be a perfect sphere. In this he was quite correct, at least at first approximation, but it was contrary to the observational science of the time, which knew nothing about gravitational attraction. Certainly it was difficult to understand how a man on the far side of the world could walk upside-down without falling into space. But the axiom of 'perfection' outweighed any such arguments so far as he was concerned; the Earth had to be spherical.

Plato, then Aristotle, took up this viewpoint about the circle axiom, and made it the basis of their cosmology. Their arguments were so convincing that the axiom was accepted as the basis of knowledge of the world system for the following seventeen centuries. Indeed, it can even be said that the effects lingered on for the following twenty-five centuries, because even in our own time we can see traces of it. When we have to choose between different theories of the universe put forward by modern researchers, there are many physicists and astrophysicists who intuitively, and spontaneously, opt for a spherical universe. Of course, their reasons are more scientific than Plato's, but it is worth re-reading what was written in the famous *Timaeus* of Plato (33B–34B):

And he gave the universe the figure which is proper and natural. For the living thing which should contain within itself all living things, that figure would be proper which contains within itself all figures whatsoever. Wherefore he turned it, as in a lathe, round and spherical, with its extremities equidistant in all directions from the centre, the figure of all figures the most perfect and most like to itself, for he deemed the like more beautiful than the unlike. To the whole he gave, on the outside round about, a surface perfectly finished and smooth, for many reasons. It had no need of eyes, for nothing visible was left outside it; nor of hearing, for there was nothing audible outside it; and there was no breath outside it requiring to be inhaled. Nor again did it need to have any organ whereby to take to itself sustenance and, when it digested, to void it again; for nothing went forth from it nor came to it from anywhere, since nothing existed but itself. It was designedly created such as to provide itself with its own sustenance out of its own waste, and to act and be acted upon wholly in itself and by itself;

for its Creator deemed that, were it self-sufficing, it would be much better than if it had need of other things. He did not think he ought uselessly to give it hands, which it could not need for taking hold of anything nor for defending itself against anything; nor yet feet, nor generally anything that serves for taking steps. For he allotted to it the motion which was proper to its bodily form, that motion of the seven motions which is most bound up with understanding and intelligence. Wherefore, turning it round in one and the same place upon itself, he made it move with circular rotation; all the other six motions he took away from it and made exempt from their wanderings. And since for this revolution it had no need of feet, he created it without legs and without feet ... he made it smooth and even and everywhere equidistant from the centre, a body whole and perfect, made up of perfect bodies.

Therefore, following Plato, the cause seemed to be understood; from all the evidence, the form of the world could only be a perfect sphere, and all movements of celestial bodies would necessarily be perfect circles. Moreover, the velocities of motion would be constant, because God would have no reason to slow down or speed up any object in the sky. Plato's successors did their best to account for the observed movements of the Sun and planets, which certainly did not fit in with the idea of steady motion in a circular orbit; their reasoning was most intriguing, and few people questioned it. According to Plato, the circle axiom could account for the entire mechanism of the universe.

Aristotle

Aristotle returned to Plato's view of a spherical universe, but he presented it so cleverly that men found that it appealed as much to their hearts as to their reason. The model that Aristotle proposed seemed to be acceptable, even though it was not in agreement with many of the opinions that had been widely held up to that time.

At the beginning of the fourth century BC, there was much discussion as to the nature of the primordial substance of the universe. Thales of Miletus claimed that water was the material cause of all things, but another philosopher of the school of Miletus, Anaximenes, maintained that air was the prime substance; Heraclitus of Ephesus argued in favour of fire. However, a considerable number of philosophers came round to the idea that earth, so common and so abundant, might be the material from which everything else was formed.

There was another problem, too. Was the universe in a state of continual change, or had God made it in his image, eternally perfect and changeless? Here, too, opinions were divided – as, indeed, they still are today. Parmenides, who lived at Elea in South Italy, made the famous pronouncement that, 'one cannot know, nor express, what is not; because that which can be thought of and that which exists are the same thing'. This led to the idea of unity. The world is One; it has neither creation, nor beginning, nor disappearance; change is nothing but an illusion. To Empedocles, who lived in South Italy about this period, the One existed only at the very start of the world, but then he was divided into Love and Hate. At one time (the time in which Empedocles lived), Hate had overcome Love. Love is therefore outside the world, but he will return to triumph over Hate; this will be the return of the One.

What could be derived from these opposite theories, and could they be merged into a more unified view of the universe? Aristotle and his school set themselves to this task. In a way, they succeeded too well; their world system and cosmology were accepted for many centuries afterwards.

Aristotle put the Earth in the centre of the system, and then, at increasing distances, came nine transparent, concentric spheres to which the planets were fixed. In order of distance from the Earth, the spheres were those of the Moon, Mercury, Venus, the Sun, Mars, Jupiter, Saturn and the stars. Finally on the outside, came the sphere of God himself. It was thought that the sphere representing God is fixed, but is responsible for making all the others rotate. The Earth, of course, is motionless, and lies at the very heart of the universe.

Unfortunately for Earthmen, the situation was very far from that of perfection. To Aristotle, man was remote from God, tainted with vulgarity, and highly imperfect; indeed, man was at the lowest possible level.

Aristotle summed up his thoughts by claiming that everything inside the sphere in which the Moon revolved (that is, the sublunar region) was subject to blemishes associated with change. On the other hand, everything outside the sphere of the Moon was changeless. Below, things were temporal; above, they were eternal.

The substances making up these two regions were far from identical. Below, in the sublunar region, there were four elements: earth, water, air and fire, together with couples of 'opposites' such as cold and heat, dryness and wetness. (This last idea was not unlike the old Chinese conception of Yin and Yang, and went back to what had been known as the 'theory of opposites'.) The four elements moved along straight lines: earth top to bottom, water bottom to top, air and water horizontally. Moreover, the four elements continually changed from the one into the other, thereby creating the sublunar region.

Beyond the sphere of the Moon came the 'ethereal' region;

1·2 Aristotle's four elements — earth, water, air and fire — were supposed to be interconvertible. Aristotle also believed in the 'theory of opposites', such as wetness and dryness.

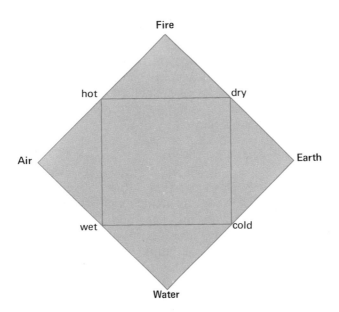

to the four elements already listed there was added a fifth, ether. The natural motion of this fifth element consisted of a perfect curve, the circle.

It is now clear how Aristotelian cosmology, based on Plato's axiom about the perfection of a circle, made up a world system which brought together the concepts of the preceding centuries. This cosmology was also broad enough to allow of different interpretations, and it was these various interpretations which dominated orthodox science for many hundreds of years afterwards.

However, there was one exception. One Greek, a 'Copernican' before his time, hit upon the right concept – which, seventeen centuries later, was to replace Aristotle's scheme of things. It was a brief spell of brilliance, and at the time it was swamped by Aristotle's immense stature. It was due to Aristarchus of Samos.

Aristarchus

Aristarchus was born in the year 310 BC, twelve years after Aristotle's death. Among his achievements was to produce a short treatise entitled *On the Sizes and Distances of the Sun and Moon*. Admittedly, the figures which he gave were inaccurate compared with those that have since been obtained by telescopic methods; but in this short book, Aristarchus was able to show his independence of vision, together with the attention to observation so necessary in true science.

From the writings left by his successors, notably Archimedes and Plutarch, we know that Aristarchus regarded the Sun, not the Earth, as being the centre of the planetary system – an idea that later philosophers did not take very seriously. Archimedes, who was almost contemporary with Aristarchus, recorded in his treatise *Of the Hour-glass*, dedicated to Gelon, King of Syracuse, that 'Aristarchus supposed that the stars and the Sun are motionless, and that the Earth moves in a circle round the Sun'. Plutarch, one hundred years after Christ, wrote in his treatise, *On the Face which is seen of the Disk of the Moon*, that 'Aristarchus thought the sky to be stationary, and that the Earth described an oblique circle, turning upon its axis'.

All this is perfectly clear. Without doubt, Aristarchus of Samos put forward the forerunner of cosmologies which became fully accepted seventeen centuries later, after the work of Copernicus, Kepler and Newton.

How, then, can we explain why Greek science, having produced so brilliant a concept, subsequently rejected it in favour of the old Aristotelian cosmology?

In modern times, there have been many attempts to explain this apparent blindness. There have been suggestions that it was more metaphysical than physical, but there may have been scientific reasons as well, and certainly it would be quite wrong to criticise the Greeks too severely. After all, it would be unfair to criticise modern scientists simply because they are bound to ignore discoveries which are certain to be made in the next few decades!

Let us first consider the viewpoint of those who consider that Aristarchus' heliocentric cosmology was 'natural and evident'. Of course it is natural to claim that the planets move round the Sun *if* one knows about gravitational attraction – because it then becomes logical in every way, just as a stone will move round in an orbit if it is swung on the end of a cord. But the idea of gravitational force is not a natural one; how can bodies attract each other across the emptiness of space, without any apparent connection? The whole idea was so far from being natural that Newton himself, in a letter to his friend Bentley dated 25 February 1692, wrote:

To suppose that a body can act upon another at a distance, across space, without the intervention of any medium whatsoever, seems to me to be so absurd that I do not believe any man capable of philosophical thought can admit such a fact.

And if the idea of attractive forces operating across empty space is dismissed as absurd, why suppose that the planets are in motion round the Sun?

Moreover, all motion is relative. It is no more natural to say that the Earth moves round the Sun than to claim that the Sun moves round the Earth. What science has to do is to work out the precise law which governs the movements of the various bodies; only when this has been done can the fundamental principles be accepted without reservation.

This, in effect, is what Aristotle tried to do. He based his ideas upon the circle axiom, and he offered a wide framework of thought which went some way toward unifying the various, often contradictory opinions about the system of the world which had been held by his predecessors.

Despite his intuition, Aristarchus could provide no absolute proof of the correctness of his theory. On the other hand, his arguments did at least have the merit of simplicity – a quality which, in our own time, meant so much to Albert Einstein. Yet, as Aristotle himself wrote: 'Faced with the most clear and simple things, men are often as blind as bats in the daytime.'

Even today, we must admit that in our quest for knowledge we still suffer from this sort of bat-like blindness.

Certainly we would be quite unjustified in condemning the blindness of the scholars of the Middle Ages simply because they struggled to see what had already been glimpsed by the genius of a solitary Greek named Aristarchus.

2 A world seeking itself

As we have seen, Aristotle believed the world system to be made up of nine transparent spheres, whose centre was the Earth. In these spheres were, in succession, the Moon, Mercury, Venus, the Sun, Mars, Jupiter, Saturn and the fixed stars; finally there came the divine sphere, symbolising the prime mover, God.

However, it was clear enough that the planets could not move in perfectly circular orbits at uniform speeds, because observation – even if carried out with very rudimentary instruments and techniques – showed that the movements of the planets were not completely smooth and regular. One fact that had to be explained was the behaviour of the planets in moving sometimes in one direction against the starry background, sometimes in the other.

Observation showed that as seen from Earth, the planets moved against the celestial sphere in paths which were shaped rather like the teeth of a saw. Some of the arcs of these saw teeth ran in one direction; then as soon as the planet reached the summit of a tooth, its movement would be backwards, or retrograde. This is easy to explain if the Sun is regarded as the central body, but in Aristotle's time, when the Earth was taken to be the centre of the whole system, there seemed to be no rational explanation at all. However, scientists can often produce a 'good' explanation in the manner of a conjurer, even if it goes against all common sense. In this case the explanation was prepared by a brilliant mathematician, Eudoxus; Aristotle had only to perfect it.

According to Eudoxus and Aristotle, the diameter line of the sphere carrying a planet is prolonged outside the sphere itself, and is fixed to another sphere, which also is turning round a diameter. Therefore, the planet has a movement round the Earth which is much more complicated than a

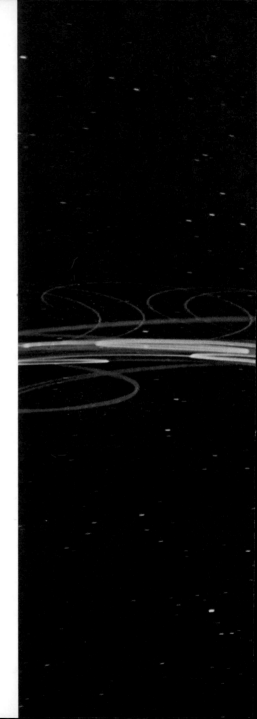

2·1 Photograph of the
apparent motions of
the planets taken in the
Munich Planetarium, where
the apparent movements
were speeded up.

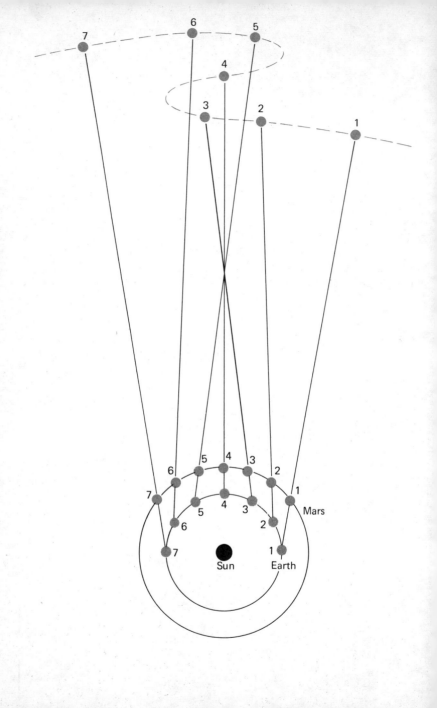

2·2 Left The diagram explains the retrograde motion of a planet, such as Mars. When the Earth 'passes' the planet, the planet seems to move in an east-west or retrograde direction against the stars for a brief period. The fact that the apparent path of Mars (top of diagram) differs from the true path (bottom of diagram) was impossible to explain so long as early astronomers believed that Mars moved round the Earth.

2 3 Below Planetary orbits according to Eudoxus and Aristotle. The diameter-line of the sphere carrying a planet is prolonged outside the sphere and is fixed to another sphere, which also turns round a diameter. The various spheres have independent axes of rotation and the path of a planet round the Earth is therefore more complex than a simple circle.

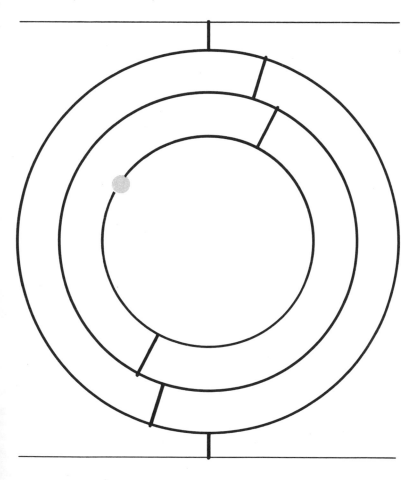

simple circle. If this is not sufficient to account for the observed motions, a third sphere can be added in the same way; if necessary, a fourth, and so on. All these various spheres have independent axes of rotation.

Eudoxus proposed twenty-seven spheres. Callippus increased the number to thirty-four. Aristotle did not hesitate to 'perfect' the cosmology by using fifty-four concentric spheres turning round the Earth. In this way, the motions of the planets were explained, at least on the larger scale. Certainly the paths were not circles, but they were worked out on a combination of circular, uniform movements, so that the fundamental 'axiom of the circle' was preserved.

Ptolemy

Then, as noted in the last chapter, came Aristarchus of Samos, who proposed to put the Sun in the centre of the world system, and thus simplify the whole theory. But men of the time were not prepared to accept a cosmology of this sort, for reasons which are difficult to appreciate today. On the other hand, it must not be forgotten that the Moon, which is the nearest body in the sky and therefore the easiest to observe, really does seem to move in a circular orbit round the Earth as centre; and it was not unreasonable to suppose that if the Moon behaved in this way, the planets should follow suit. Without a knowledge of the laws of gravitation, it was unquestionably hard for the Greeks to believe that the Moon, with its circular and almost perfectly uniform path, does not provide a good example of the principles of motion of all celestial bodies.

Yet there was one observational fact which Aristotle's spheres could not explain, no matter how many of them there might be. The planets seem to approach and recede from the Earth. If they moved on spheres concentric with the Earth, as Aristotle proposed, their distances should obviously remain constant, but in actual fact their apparent brilliancies change considerably, so that their distances from the Earth must change too. It was known that this was true even for the Moon. Eclipses of the Sun, due to the Moon passing between the Sun and the Earth, are sometimes total and sometimes annular; in the latter case, the Moon must clearly be further from the Earth than at the time of a total eclipse.

All this was well known in the time of Plato and Aristotle, but no explanation was forthcoming. It was not for another five centuries that the problem was really tackled. The astronomer responsible was Ptolemy of Alexandria (Claudius Ptolemaeus), who lived during the second century AD.

Like his predecessors, Ptolemy accepted the circle axiom. He believed this principle to be unassailable, and in his *Almagest* (III, chapter 2) he explained it clearly:

> We believe that the goal that the astronomer ought to aim at is the following: to show that the phenomena of the heavens are reproduced by circular, uniform motions.

Instead of supposing that the planets were anchored in solid spheres, Ptolemy believed them to be fixed to wheels. To visualise what he meant, it will be helpful to picture the fairground great wheel; the planets can then be represented by the passengers' chairs. Each chair has two independent movements. It turns around the axis which is fixed to the rim of the wheel; this axis is itself carried round as the wheel revolves, so that it describes a circle around the hub. The diagram shows what happens when, with the wheel rotating, the passenger's chair performs quick turns around its axis of

2·4 Left At the time of Plato and Aristotle it was known that eclipses of the Sun were sometimes annular and sometimes total (*left*) but an explanation of this had to wait for Ptolemy five centuries later.
2·5 Below Solar eclipses are due to the Moon passing between Sun and Earth. In an annular eclipse, the Moon is further from the Earth than during a total eclipse.

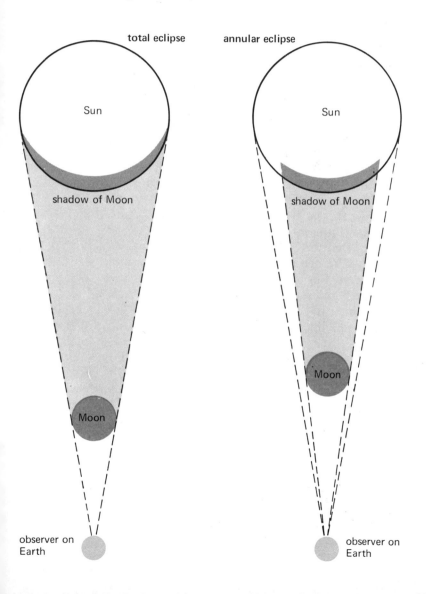

suspension, while the wheel revolves at a slower rate.

This solution of Ptolemy's may be said to be a step forward, because it is essentially the same, in principle, as modern ideas of the movements of the planets round the Sun – though the underlying theories are very different. And it is true that Ptolemy did succeed in accounting for the apparent motions of the planets in the sky; also, he explained the changing distances between the planets and the Earth.

Ptolemy's scheme has often been ridiculed in modern times, but in my view there is no justification for anything of the sort. If the observer regards the Earth as the central body (and in Greek times, there was no valid reason to think otherwise), the Sun and planets would indeed move very much as Ptolemy described.

Admittedly, there were some problems which were not solved. The actual orbits of the planets round the Sun are not circles, but ellipses; Mercury, which is relatively close to the Earth and was accurately observed in ancient times, has a path which is notably eccentric. But these details are refinements, and the great wheel model did account for the basic motions, even though it had to be considerably elaborated. As a first approximation, seven wheels would take care of the seven bodies thought to be moving round the Earth – the Sun, the Moon and the five planets; but Ptolemy wanted to explain all the various irregularities, and he proposed a system which included thirty-nine wheels. It is difficult to criticise him for this. After all, modern physicists, arguing about the mysteries of elementary particles, do not hesitate to use very complex models which involve strange phenomena – isotopic spin, for example – which, all in all, are hardly more real than Ptolemy's wheels. Inevitably, rather complicated models are sometimes necessary to explain the observed

2·6 A geocentric theory of the universe. A planet moves in a small circle, or epicycle, the centre of which (the deferent) itself moves round the Earth in a perfect circle. In this way Ptolemy succeeded in accounting for the apparent paths of the planets. Later, Ptolemy added more and more epicycles to account for all the planetary paths.

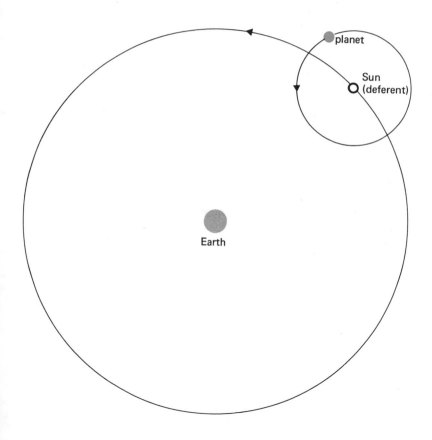

facts, but the complexities are usually jettisoned as soon as is possible. To progress is, essentially, to simplify.

Moreover, it is worth noting that since all movements are relative (as Einstein has shown so clearly), there was originally no reason to choose a system of reference based upon a central Sun instead of a central Earth. Ptolemy accounted for the movements, but he was unable to say why the planets moved as they did. He could not be expected to provide an explanation, because he knew nothing about gravitational attraction. Aristarchus had been equally ignorant about gravity, even though he had believed that the planets turn round the Sun; therefore, it is not fair to regard Ptolemy's cosmology as a retrograde step compared with that of Aristarchus. Many centuries separate Aristarchus from Copernicus, which shows that it takes mankind a long time to understand the mechanics of the immense universe.

One thing is worth stressing. All the ancient scientists – Ptolemy; further in the past, Aristotle and Pythagoras; even further back, the Egyptians and the Chinese – appreciated the importance of the Sun, and ranked it above all other celestial bodies. Indeed, the Sun was regarded as the God of the universe. Cicero, 106–43 BC in the *The Republic*, even wrote: 'The Sun, master, prince and guide of all other heavenly bodies, unique ruler of the universe, is so great that its brilliant light penetrates all....'

Pliny, Plutarch, Julian the Apostate, and Macrobius, in the first centuries AD, repeated the same sentiments in various forms; all of them believed the Sun to play an essential role in the universe. It was left to Newton to suppose that the Sun is nothing more than a star. But the ancient philosophers had made great strides, and they had laid the foundations of modern knowledge.

Saint Augustine

Something has already been said about man's habit of
increasing his knowledge by theoretical and observational
methods. This has happened often enough in the past, though
nowadays, of course, theory and observation go hand in
hand. One generation would concentrate chiefly upon *a
priori* hypotheses; the next would rely upon observation.
Thus Pythagoras, the observer, was followed by Plato, with
his axiom of the circle; then came Aristotle, who preached
observation (even if he did not always carry it out). During
the last few centuries BC, the emphasis was upon perfecting
Aristotle's cosmology by adding more and more spheres,
even though the changing distances of the planets made the
whole scheme implausible. Next came Ptolemy, who tried to
reconcile cosmology with the actual observed movements of
the planets. After Ptolemy there came a new phase, a curious
mixture of science and religion. Between the second and ninth
centuries AD the general view was that there was little point in
observing the phenomena of nature, since all essential truths
were to be found in the human mind and in the sacred books
concerning God and the destiny of man.

This attitude was particularly marked in the case of St
Augustine, who lived toward the end of the fourth century.
In his *Confessions*, 10th Book, he wrote, following a section
about the sins of the flesh:

There is another form of temptation, even more fraught with
danger. This is the disease of curiosity ... it is this which drives us
on to try to discover the secrets of nature, those secrets which are
beyond our understanding, which can avail us nothing, and which
men should not wish to learn ... In this immense forest, full of
pitfalls and perils, I have drawn myself back, and pulled myself
away from these thorns. In the midst of all these things which float

unceasingly around me in my everyday life, I am never surprised at any of them, and never captivated by my genuine desire to study them ... I no longer dream of the stars.

This does not mean that St Augustine did not attempt to formulate a cosmology – in other words, a model of the system of the world. But this cosmology was drawn entirely from his own feelings. He was not in the least concerned with finding out whether nature would strengthen or weaken the image of the world system he had formed in his mind.

We must pause for a moment to examine this concept – first because successive generations, over the centuries, took the same attitude even though their eyes, ears and brains were as good as ours. We find their outlook surprising today, but it was almost universal then. Also, men can increase their knowledge only by building upon the ideas drawn up by their predecessors, so that in a way our own knowledge has been built upon this curious outlook so common in the past.

An outlook such as that of St Augustine brings home to us, very strongly, that our knowledge is not to be regarded as absolute; it is always relative to ourselves. Nothing is truly black or truly white; it may be taken as being either black or white according to the viewpoint from which we look at it. The principle applies equally to what we may call more 'human' problems, dealing with life, death, God, and the universe considered as a whole.

What is the universe like, in those remote regions which we can never observe? This is a question that St Augustine might well have asked himself: Who can say that my opinion of nature, based on what I can see close to me, might not change completely – even to the point of seeing white things become black – if I could find out more of the secrets which are completely beyond my present knowledge? I believe that

the Sun moves round the Earth, he would continue, but I might
hold a different view if I knew a little more about the system
of the world. This quest for knowledge is like looking into a
bottomless well. There are always more and more facts about
which I am ignorant; it is thus today, it will be thus tomorrow, in a hundred years or in a thousand years. So be it.
The true values are to be found in my own mind; I must take
hold of them before physical hypotheses, apparently necessary even if always provisional, can obscure my view and
mislead me with regard to the course which I must follow.

Successive generations of scholars, over the centuries, have
given to the outer world an image of a place which favours
their contemplations. (After all, there is no reason why they
should not be called scholars.) For prayer, why do we go to a
church, a temple or a mosque? – simply because the conditions there are favourable for direct, intuitive communica-
tion. If we are to understand the meaning of human existence,
this is surely essential.

Then, St Augustine might have said – just as the scholars
of the Middle Ages might have said – then shut your eyes,
and picture the world as a place offered to man for meditation. Why should the Earth be round? Places of worship
certainly do not have so strange a form. So let us opt for a
flat Earth. If we must have a definite shape for the Earth, let
us look for details in the sacred books, particularly in the
Bible. In Genesis 1, 6–7, we read: 'And God said, Let there
be a firmament in the midst of the waters, and let it divide the
waters from the waters. And God made the firmament, and
divided the waters which were under the firmament from
those which were above the firmament'.

St Augustine, then, built his cosmology upon thoughts
such as these. The firmament was surrounded by water; the

Earth was flat; there was water above the sky, and more water below the Earth.

Two centuries after St Augustine, a monk named Cosmas set out to perfect this world model. In the first of his twelve books, entitled *Against those who, even though professing Christianity, believe and imagine, as heathens do, that the heavens are spherical*, Cosmas claims that the world is rectangular, and twice as long as it is broad. The Earth is a sort of flat island in the middle of the world; it is surrounded on all sides by water. Some distance away there is a second Earth; this is paradise. The universe is enclosed by high vertical walls, and the whole of the universe is a half cylinder. The stars are put in their places by angels. An enormous mountain is situated at one of the ends of the Earth; the stars disappear when the angels push them behind this mountain, so hiding them from man's eyes.

All this sounds rather like a child's fairy story, and certainly it created an impression of child-like safety and comfort. Man knew where he came from, where he was going, and what he was; he had his whole life to think about other problems, and to pave the way for his descendants. This kind of attitude was hardly calculated to lead to revolutionary new techniques, but man has not been put on Earth merely to work out new techniques. Humanity has a fundamental aim, just as a gardener cultivates a rose bush so that later on it can produce flowers.

Aquinas

At the start of our own era, there was a new shift to the type of knowledge characteristic of the period of Greek brilliance, between Pythagoras and Aristarchus of Samos.

Ancient Greek learning, written down for our benefit in the manuscripts of Plato and Aristotle, was, happily, not lost; it was carefully preserved by Arab science. Much of the activity of this Arab science, during the first centuries AD, consisted of translating Euclid, Ptolemy, Galen and others, including Aristotle, into Arabic. These Arab texts still exist, though the Greek originals have never been recovered.

The Arabs regarded Greek science as representing absolute knowledge, so that there was no need to try to perfect it further. For the Arabs, then, science became largely a matter of hunting for manuscripts – as though these old manuscripts contained the secrets of the entire universe. All the same, this work of translation, and consideration of the views of the ancient philosophers, ended in the development of an original scientific spirit, and a glut of new researches. The great translators flourished during the ninth and tenth centuries; many of them also wrote texts based on their own personal researches. In Cairo, an Academy of Sciences soon appeared. Astronomy, physics, and (in particular) optics were subjects in which very valuable work was carried out.

The contact between the Latin West and the Moslem West came by way of military invasions. In 1136 Cordoba, the cultural capital of the Moslem West, was taken by Ferdinand III, King of Castile. In 1258 Baghdad, another very active scientific focal point, fell under Mongol control. Arab knowledge, in which Greek learning had been so carefully preserved, came slowly back toward Europe by way of Spain.

It was in Spain that Aristotle's works were translated into Latin. This was at the beginning of the twelfth century; by then Aristotle's *Physics*, Ptolemy's *Almagest* and Euclid's *Elements* had once more become official scientific texts, and the future progress in Europe was founded upon them.

The Earth was again regarded as spherical, even if the existence of people living in the antipodes, in the regions below the seas, was still doubted. There was renewed interest in the hypotheses of Ptolemy – and even in those of Aristarchus. Above all, scientists began once more to care about practical observation, and there was frequent reference to Aristotle's words from *De Coelo: De Generatione et Corruptione*, in which he underlined the importance of experience as against *a priori* conclusions:

It is easy to distinguish between those who reason according to the facts, and those who reason according to ideas. The principles of all the sciences are derived from experience; thus the principles of astronomy are drawn from observations of the celestial bodies.

This thirteenth-century renaissance of a desire for knowledge of nature seems, in the main, to have been the rediscovery of the power of methods of reasoning. During the preceding centuries, reason had been completely subordinate to faith; to know was to believe. Now, it was again realised that knowledge can be independent of belief – and that true knowledge is due to authentic reasoning, capable of building upon hypotheses drawn by observations of natural phenomena instead of theological texts.

In the middle of the thirteenth century, Thomas Aquinas provided a particularly brilliant example of the way in which to use reason to tackle the most difficult problems. Admittedly, appreciation of God remained an essential preoccupation in all knowledge, especially knowledge of natural phenomena. But even with regard to knowledge of God, Thomas Aquinas did not rely solely upon his intuition; he had recourse to reason. The themes developed from scientific axioms, and depending entirely upon the foundations provided by these axioms, led to methods which were later to become those of modern science. Thomas Aquinas expressed himself in terms of power and deed, or matter and form, or body and spirit; no doubt the terms were metaphysical compared with those of today, but they nevertheless amounted to an attempt to analyse the phenomena of nature by pure observation, and also to analyse the working of the human spirit. As Stendhal remarked later in his *Pensées diverses*: 'We see things that our head describes to us, and therefore it is necessary to know this head.' This sort of attitude was one of the first signs of a new method of increasing knowledge – a method which was ambitious as well as precise. Man came to realise that even though he is insignificant in the universe, his reasoning powers combined with observation are capable of allowing him to probe the mechanism of the whole universe. Moreover, men found that the world is not a place where things occur according to the whims of angels. It is a world controlled by laws, and above all by the Sun.

3 A world subordinate to the Sun Copernicus

In one way, the history of cosmology resembles the history of man himself. There are periods which stand out, during which individual men put forward ideas which are confirmed and improved by following generations.

There was an abrupt spread of knowledge during Pericles' century, and the glow lasted for the following two centuries, suddenly shining forth once more with Aristarchus of Samos in the third century BC. Then, for the next fifteen centuries, there was almost no progress in the fundamental problem of understanding the world and the universe which surrounds it. During all this time scholars tended to look inward; man thought himself to be the sole source of knowledge. And though the scientists did not lose sight of the models proposed by Aristotle, Heraclitus and Aristarchus, these models were regarded as coarse constructions which dealt only with outward appearances. After all, it is true that Aristotle's spheres and Ptolemy's wheels were hardly convincing enough to persuade thinkers that they represented reality. Moreover, the scholars of the time had as much spirit as ourselves; if they lacked models satisfactory to their outlook, they took refuge in a view of the world system which was inward looking and intuitive.

Unfortunately, this intuitive view was closer to art (even religious art) than to science. There was no language capable of describing the mechanisms of nature to anyone who was prepared to open his eyes and look into the outer world.

Slowly this attitude passed away. Eyes were opened; reason returned as a means of increasing knowledge; there came a period of two centuries during which matters moved forward with startling speed. This was the time of Copernicus (1473–1543), Galileo (1564–1642), Kepler (1571–1630), Descartes (1596–1650) and Newton (1642–1727).

The best method here will be to discuss the lives and works of these five men in chronological order. We begin with the first of these pioneers, Nicolaus Copernicus.

The work of Copernicus is generally regarded as the first spark which started the great scientific movement of the Renaissance. To his contemporaries, the vital point was that in the Copernican view, the Earth no longer occupied the centre of the world system; it was simply a planet which, like the five others, moved round the Sun. However, there was a long period during which almost nobody knew about Copernicus' results. Their publication was delayed, and the work was known only through some manuscript documents which were circulated almost surreptitiously.

When Copernicus was almost seventy years old, he at last made up his mind – but only at the urging of a young scholar of twenty-five, Rhaeticus, who took upon himself the task of seeing the work through the press. It was high time to take action, both for the world and for Copernicus himself; indeed, he was on his death-bed when the first copy of the book was brought to him straight from the printers.

Undoubtedly this book was the most valuable of Copernicus' works. But although his name is known to everyone today, it did not become so for a long time after the publication of his book. The first edition, of which a thousand copies were produced in 1543, was never sold out. On the other hand, the manual *Doctrines of Physics* by Melancthon, which was written solely to refute Copernicus' ideas, went to seventeen editions before the first reprint of Copernicus' *De Revolutionibus*. Copernicus hesitated for a long time before giving his work to scientific history; history repaid him by hesitating for a long time before allowing Copernicus to enter its great pages.

Non parem Pauli gratiam requiro
Veniam Petri neq, posco, sed quam
In crucis ligno dederas latroni
Sedulus oro.

3·2 Below Copernicus' geometry room at Cracow, where he pursued his early astronomical studies. Many traditional Euclidean diagrams are shown on the wall.

3·3 Right Copernicus' workroom at Frauenburg, where he applied himself to the study of astronomy and mathematics. The quadrant could be used to measure the position of the Sun.

Copernicus was never poor. On the death of his father in 1484 (when the young Copernicus was ten years old) he was brought up by his uncle Lucas, who was a bishop by profession. Lucas gave Copernicus a sound education, and when Copernicus was twenty-two years of age, he gave him the living of Chanoine at the Cathedral of Frauenburg. It was a task which left him considerable spare time. Copernicus accepted the income which went with the living, but during the fifteen years following his appointment he was almost never to be found in Frauenburg; he was following up his studies in different universities.

At this period, advanced studies went on for a very long time, and anyone who set out to become a scholar had to devote virtually his full attention to it. Copernicus' training lasted for fourteen years, during which time he learned philosophy, law, mathematics, medicine, astronomy and Greek. At the age of thirty he was at last given the degree of Doctor of Law from the University of Ferrara; he had studied successively at the universities of Cracow, Bologna and Padua. In 1506 Canon Copernicus decided that his studies were complete, and he returned to Prussia.

His attention was not yet fully concentrated upon astronomy. He was still Canon of Frauenburg, but he granted himself leave of absence to become official doctor to his uncle, Bishop Lucas, at the Castle of Heilsberg. His duties included taking care not only of his uncle, but indeed of the whole region of Ermland in which the castle lay. We can find traces of the care which Copernicus gave to all the people who called on him for consultation, some of whom came from great distances. It must be said that Copernicus dispensed curious kinds of medicines. In his writings we note certain recipes that he considered particularly curative, for example, Armenian sponge, cedarwood, scrapings of ivory, crocus and saffron, camomile and vinegar, red and blue hyacinths, powder of deer's hearts, a beetle, unicorn's horn, red coral, and gold, silver and sugar. Happily, our canon-physician seems to have shown more discernment in astronomy!

The Commentariolus

At last Copernicus was able to concentrate upon astronomy. His duties as a doctor were hardly more exacting than those as a canon, and they left Copernicus plenty of time to follow up his personal interests. During his leisure time, he pondered upon the problem that had intrigued him so deeply for so long: that of the movements of the planets.

In the course of his studies, he had learned about the ideas put forward by Aristarchus of Samos almost twenty centuries earlier – ideas according to which the planets move round the Sun, not round the Earth. This system impressed him at once by its simplicity; he had an instinctive feeling that the basis of the idea must be correct, and he began to examine the arguments which could be brought forward to support it.

He was thirty-six years old when he wrote down his findings in a short manuscript, which at first he certainly meant to publish. The manuscript is known as *The Commentariolus*, which may be translated as *Résumé*. Essentially, it was a summary in which the broad scheme of the future Copernican synthesis made its first appearance. However, it was not printed; it remained as an outline, and it was circulated only among Copernicus' close friends.

When he was not working upon the theory of planetary motions, Copernicus dealt with subjects other than astro-

3·4 The Copernican theory as presented in *De Revolutionibus Orbium Coelestium.* The Earth no longer lay at the centre; it was now one of a number of planets which moved round the Sun. Beyond Saturn comes the sphere of the fixed stars.

nomy, and devoted some of his leisure time to translations of ancient writers of no great interest – for example, he translated the Letters of Simocatta, a Byzantine historian of the seventh century. The Latin version, containing eighty-five letters, made up a thick volume. It is questionable whether the work may be said to have dealt chiefly with moral, pastoral or romantic themes. Consider Letter 84, from Chrysippe to Sosipater:

You are caught in the net of love, Sosipater; you love Anthuse. Well deserving of praise are the eyes which amorously follow a beautiful virgin! Do not pity yourself because you have been conquered by love; the delights of love are greater than the penalties. Though tears come from chagrin, those of love are sweet, because they are a mixture of joy and pleasure. The gods of love bring their delights together with sadness; Venus surrounds herself with many passions.

We have come a long way from astronomy – except, perhaps, for the mention of Venus!

The Book of Revolutions

In 1530 Copernicus was already fifty-seven years old. He had returned to his studies of the movements of the planets, and he had written a book which was more complete than the *Commentariolus;* he called it *De Revolutionibus Orbium Coelestium (The Book of Revolutions).* In the preface to the manuscript of his book, which he proposed to dedicate to Pope Paul III, Copernicus states that he had delayed publication because of the fear of ridicule. He does indeed appear strangely timid for a man who sets out to change the whole accepted view of the universe. Copernicus writes:

I can presume, Holy Father, that certain people, learning that in

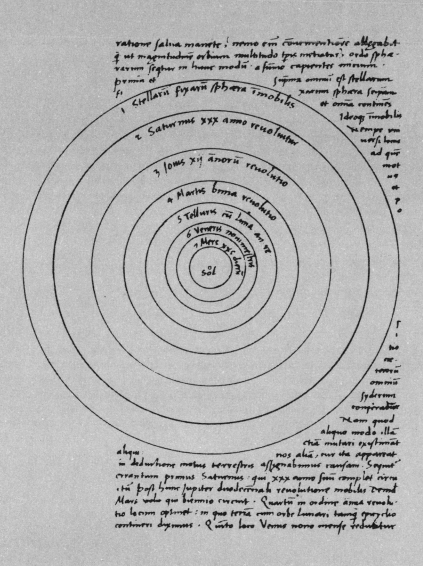

ratione salua manente : nemo enim commodiorem allegabit
q̃ ut magnitudinis orbium multitudo tp̃e mutatur: ordo sphæ-
rarum sequitur in hunc modum · a summo capientes initium ·
prima et   suprema omnium est stellarum
si   rarum sphæra seipam
                                  et omnia continens
                                  Ideoq̃ immobilis
1 Stellarum fixarum sphæra immobilis   sc̃m se locus
                                  ad qu̅
2 Saturnus XXX anno reuoluitur   mot
                                  us
                                  et
                                  po
3 Iouis XII anoru̅ reuolutio

4 Martis bima reuolutio
5 Telluris cu̅ Luna an. re
6 Veneris nonime̅ltri
7 Merc XXX dies
Sol

                                  s·
                                  no
                                  ce-
                                  teroru̅
                                  omniu̅
                                  syderum
                                  conspiratur
                                  Nam quod
                                  aliquo modo illa̅
                                  ctiã mutari existimat
aliqui: nos aliã, cur ita appareat
in deductione motus terrestris assignabimus causam. Sequitur
crantum primus Saturnus: qui XXX anno suo complet circu-
itu̅ post hunc Iupiter duodecennali reuolutione mobilis Deind
Mars uolo qui biennio circuit · Quartu̅ in ordine a̅nua reuoluti-
tio locum optinet: in quo terra cum orbe Lunari tanq̃ epicyclio
continetur dicimus · Quinto loco Venus nono mense reducitur

this book *The Revolutions of the Celestial Bodies* I attribute certain movements to the Earth, will write that for holding such opinions I ought immediately to be jeered at and expelled. For a long time I have asked myself whether I ought to publish my proofs of the movement of the Earth, or whether it would be more valuable to follow the example of the Pythagoreans and others, who were willing to pass their philosophical mysteries on only to their friends and intimates by word of mouth rather than by writing ... Yet in considering this subject, the scorn which will be directed at me for my new and (apparently) absurd opinions has failed to make me abandon my project.

And yet Copernicus had still not made up his mind. Time passed by, and he published nothing. However, he began to discuss his ideas in certain restricted circles. In 1536, when he was sixty-three years old, he had a letter from Nicolas Schönberg, Cardinal of Capua, persuading him to publish. This letter was inserted by Copernicus in the preface to *The Book of Revolutions*. Schönberg wrote:

I have learned that you know not only the groundwork of the ancient mathematical doctrines, but that you have created a new theory of the universe, according to which the Earth is in motion and it is the Sun which occupies the fundamental, and therefore cardinal, position....This is why, wise scholar, without being obtrusive, I urgently entreat you to communicate your discovery to the world of learning.

It is very likely that the Cardinal wrote to Copernicus not on his own initiative, but on the insistence of the Pope, Clement VII, who was keenly interested in astronomy. If so, it should have shown Copernicus that he had nothing to fear from the religious community for his unorthodox interpretation of the biblical texts – texts which showed (at least according to the accepted interpretation given to them at the time) that the Earth is immovable in the centre of the world system.

But be this as it may, Copernicus still withheld publication.

Rhaeticus

It was then that Rhaeticus arrived. He was a young professor of mathematics, twenty-four years of age, holder of the chair of mathematics and astronomy at the University of Wittenberg. He owed his nomination to Melancthon, then a well-known astronomer and later a violent opponent of Copernicanism. Rhaeticus did not share Melancthon's views. On the contrary, he had always preferred the heliocentric theory proposed so long before by Aristarchus of Samos; presumably Melancthon chose to ignore this quirk when he nominated Rhaeticus for his post at Wittenberg. Rhaeticus had heard talk of Nicolaus Copernicus who, it was said, had revived and refined Aristarchus' ideas. And in the spring of 1539, when Copernicus was sixty-six years old and had only another four years to live, Rhaeticus requested leave of absence from his university so that he could go to consult the man whom he already regarded as his master.

Rhaeticus arrived at Frauenburg, bringing presents for Copernicus. He met the ageing Canon, and at once brought up the project which was so dear to his heart: to publish the work which Copernicus guarded and concealed, and to make his theories of the universe known throughout the world.

Copernicus, however much he may have been influenced by this fiery disciple, still hesitated. Then he suggested that he might publish only some numerical tables concerning the movements of the planets – based upon his cosmology, certainly, but without giving the explanations which would justify the figures, and without saying anything about the motion of the Earth.

Rhaeticus dismissed this solution out of hand, because it would remove the essential part of the whole argument. At last he found a way out of the dilemma. He suggested that he should himself write an account of Copernicus' manuscript; it would be called the *Narratio Prima* (First Account), and it at least could be published. Copernicus agreed, but insisted that his own name should not be mentioned anywhere in the book.

Rhaeticus set to work. In ten weeks he had produced a small book of seventy-six quarto pages, in which he summarised the fundamentals of Copernicus' cosmology – but without naming Copernicus even once. In the final printing, the title of the work as put out by Rhaeticus was *A First Account of the Book of Revolutions of the very learned and very excellent mathematician, the Reverend Doctor Nicolaus of Torun, Canon of Ermland, by a young student of mathematics.*

The work consisted of a really brilliant explanation of Copernicus' ideas. Copernicus' own manuscript amounted to six enormous books, covering the whole field of astronomy, with a great number of numerical tables. Rhaeticus summarised all this perfectly; within a month he had drawn up a very clear text – clearer, indeed, than that of the master himself. The *Narratio Prima* was printed in February 1540, about six months after the first meeting between Copernicus and Rhaeticus. At once Rhaeticus sent copies to well-known astronomers and other eminent persons; Melancthon received one, Duke Albert of Prussia another. Gassarus, a mathematician friend of Rhaeticus, was immediately enthusiastic when he read the book, and arranged for another edition to be printed at Basle (the first edition had been produced in Danzig).

All the same, Rhaeticus' book did not succeed in satisfying the scholars who were interested in Copernicanism. What they wanted was the complete book written by the master himself. Again Copernicus was asked to publish. Somewhat reassured by the generally favourable reception given to the *Narratio Prima*, he did, at last, consider the idea more seriously. He felt that it would be necessary to fix a further meeting with Rhaeticus, who had returned to his class at the University of Wittenberg. In the spring, when the classes had been completed, Rhaeticus duly came back to see Copernicus.

By now it was 1540. Rhaeticus did not go back to Wittenberg during the following university year; he stayed near Copernicus, re-reading the manuscript, re-writing the tables clearly, making necessary corrections – in short, putting the manuscript into publication form. By the time he had finished, he had re-copied almost all the book. In August 1541 he had to return to Wittenberg to carry on with his teaching; and when he came back to Frauenburg in May 1542, he found that Copernicus had done nothing to speed up publication of the book. Copernicus was now sixty-nine, and had only one more year to live.

In haste, Rhaeticus took the manuscript to Nürnberg, where he expected to be able to arrange for publication. Unfortunately, his University chose this moment to take disciplinary action against him, not because of his prolonged absence with Copernicus, but because exception had been taken to his morals. He moved from Wittenberg to Leipzig, and in November 1542, by which time the book had still not been printed, he began his classes at Leipzig. The task of printing the book, and writing a preface to it, was entrusted

3·5 Chart showing the life-spans of the great cosmologists of the sixteenth, seventeenth and eighteenth centuries.

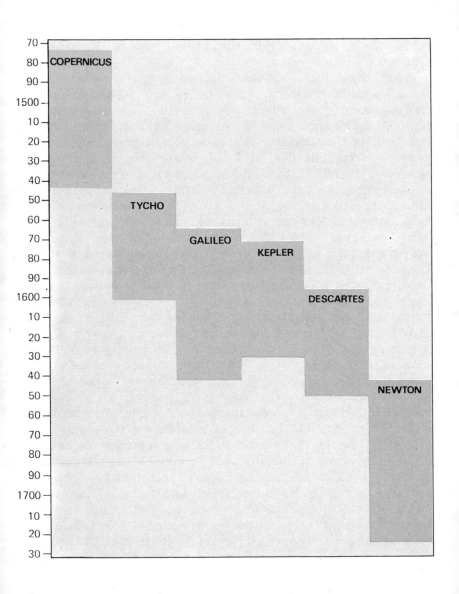

to his friend Osiander, who was known as an eminent theologian.

Copernicus had already become ill, and the winter of 1542–3 was destined to be his last. A few hours before his death, in May 1543, his book concerning *The Revolutions of the Celestial Bodies* was given to him, but he was unable to read it. Happily, he never knew about Osiander's preface, which was headed 'To the reader, concerning the hypothesis given in this book'. Osiander said, in effect, that the theory proposed in the book need not be taken too seriously, and that the ideas 'need not be true, or even probable ... the work contains, moreover, absurdities which need not be referred to here'.

Truly, this is a sad epitaph for a man who – even if, to some extent, in spite of himself – has become the symbol of objective science, a science in which man has at last left the centre of the stage to become, more justifiably, a humble participant.

The Copernican world system

It is now time to discuss the Copernican cosmology, so that we can see how it was able to exert so profound an effect upon the ideas of men of succeeding generations.

The whole work consisted of six books, but in the first twenty pages Copernicus put forward the fundamental ideas. The rest of the work consisted only of the applications of these basic principles.

However, the first twenty pages were of paramount importance. In them, Copernicus explained clearly and without ambiguity that the Sun lies at rest at the centre of the spheres of the fixed stars. Around the Sun turn the planets, moving in paths which are sensibly circular and sensibly in the same plane. In order of distance from the Sun, the planets are Mercury, Venus, the Earth, Mars, Jupiter and Saturn. The Moon turns around the Earth, and is associated with the Earth in our circular motion round the Sun. Moreover, the Earth turns on itself; this accounts for the regular alternation of day and night, since the Earth is lit up by the Sun. Minor irregularities in these various motions are due to oscillations of the Earth's axis.

The scheme was more precise than any that had been proposed before – and, incidentally, certainly more precise than many of the views widely current in our own century!

On the other hand, Copernicus retained a good many of the ancient prejudices. The celebrated dogma of the circle, in particular, seemed to him to be absolutely valid. Copernicus was not an observer, and therefore it was not by observation that he had concluded that the planets must have circular orbits; he believed implicitly that the paths must be perfectly circular, since only the circle is the perfect curve. In fact, Copernicus was blindly following the ancient tradition. His lack of freedom of spirit when considering the ancients (though, fortunately, he was much more free when dealing with his contemporaries) is well shown in the following text, which he wrote near the end of his life:

We must agree to follow strictly the methods of the ancients, and to keep to their observations, which we must regard as a Testament. And to those who think that they are not worthy of trust in this respect, the doors of our science will certainly be closed.

This attitude is a long way away from that of Descartes; not for another century did there come a new manner of looking at the universe. During that century two brilliant men lived and worked, each responsible for fundamental advances in knowledge. These two men were Kepler and Galileo.

4 A world subject to laws Kepler

Kepler and Galileo mark the real transition between the world system of the ancients and that of modern times. As we have noted, Copernicus was full of respect for the authority of the ancients. Even when he proposed his heliocentric system, he did not genuinely revive the idea put forward by Aristarchus of Samos. Instead, he was very circumspect, and took pains to assure himself that his cosmological model was in accord with the Platonic dogma of uniform motion in a circle.

Kepler, like Galileo, took a somewhat different view. He wanted to retain some links with the past, but at the same time he was determined to look forwards.

Kepler's efforts to reconcile his theories with those of ancient times are summed up by his acceptance of Plato's five regular solids as essential entities in the world system. All his life he believed that harmony must be the fundamental quality which regulates the form and the movements of all the bodies in the universe; his last work was called *The Harmony of the World*. Pythagoras was never far from Kepler's thoughts. Yet Kepler was basically modern; this will be shown in the pages which follow, dealing with the importance which he attached to a really exact agreement between theory and observation. There will always be two images which can be conjured up in Kepler's picture of the universe: an abstract image, which is beautiful, harmonious and spiritually satisfying – this is the image taken to be fundamental, and so essential to our knowledge; in the background we must then do our best to demonstrate the mechanism of this ideal image, and to make it agree with observational data. Since this cannot be done successfully, we must construct another model, a second world image, which must be worked out on a mathematical basis. Only this second image can be passed down to succeeding generations, and only this second image – which Kepler thought was less important than the first – can be used today as a foundation for building up knowledge.

Kepler was one of the last great scientists who deliberately set out to produce a world system which was both beautiful (because it was the reflection of the creator) and exact (because it had to serve man). To Kepler, the world was rather like a work of art, such as a beautiful painting. The overall impression gave him his first image, and only then did he start to analyse its details, much as a painter selects his colours or lights his canvas. To Kepler, there was no great schism between art, religion and science. Knowledge could still build up a world system which was harmonious; God was not totally absent from it, but, all the same, the image was scientific.

Galileo, in Italy, played a role similar to that of Kepler in Germany. Like Kepler, he had one foot in the past and the other in the future. All his life, Galileo defended the idea of uniform circular motion of the planets round the Sun; like Copernicus, he kept to the ancient dogma, and in this respect he even opposed Kepler. Yet Galileo was modern enough with regard to his care in making exact observations; otherwise he could never have discovered the four satellites of Jupiter and the phases of Venus. Like Kepler, he was one of the real founders of the Newtonian universe; he worked out the essential laws of dynamics, and in particular he showed that a body moving in a vacuum at a certain velocity will continue to move at the same velocity unless acted upon by any external force. This was a departure from Aristotelian theory, according to which the body would eventually come to a stop – from which it followed that the planets would have to be continually pushed along by some mysterious force.

Cosmic secrets

With this introduction, let us come back in rather more detail to Kepler and his cosmology.

Kepler was born on 16 May 1571, in Germany – to strange parents who certainly did not seem destined to set him upon a scientific career. His father was a sort of adventurer, who became a mercenary concerned with fighting the Protestant rebels in the Low Countries, first on behalf of the German Emperor and then for the Duke of Alba. He was threatened with death by hanging, and when Johann was seventeen years old his father disappeared for good. Possibly he enlisted in the Neapolitan fleet. In any case, nothing more was ever heard of him.

Johann's mother seems to have been just as erratic. She had been brought up by her aunt, who had subsequently been burned at the stake for witchcraft. From this aunt, she had acquired a taste for sorcery; she grew herbs, prepared magic potions, and tried to influence people around her. This, then, was her own particular brand of science. She later escaped her aunt's fate only because of the intervention of Johann Kepler, by then a well known and highly respected astronomer.

One way and another, Johann's family was poor and disunited. Fortunately, Germany at this time had already adopted something like the modern western system of education. There were scholarship grants for children who were poor but gifted, and the young Johann came into this category. He received a good education, first at school, then at a seminary and later at the University of Tübingen, where, at the age of twenty, he obtained his diploma. Next he registered at the Faculty of Theology in the same city, and worked there for four years. However, he did not complete the theological course, which ought to have lasted for at least seven years. When he was twenty-four years old, he was offered the post of Professor of Mathematics at the Protestant University of Grätz. Kepler hesitated, but the post ensured that he would be financially independent, and finally he accepted.

It can hardly be said that Kepler was successful as a professor. In the first year he had a dozen students in his class, and in the second year none at all. Oddly enough, the directors of the university continued to pay his salary without question; they simply allotted him a small supplementary course on Virgil and rhetoric.

It was at this time that Kepler began to gain a certain reputation in astrology. The duties of a professor of mathematics at the university included the preparation of annual almanacs containing astrological predictions. In his first almanac, Kepler predicted a major cold spell and a Turkish invasion, both of which duly occurred.

To be fair, it may well be that Kepler did not take his predictions very seriously. All the same, he regarded astrology as an exact science. Moreover, he was always preoccupied with the problem of his own destiny. He thought that the destinies of men are predetermined, and cannot be altered by free will. Many people of today would agree with this view, although astrological science is still waiting for a Newton or an Einstein.

Kepler's investigations of cosmology began in his university days. He was a firm supporter of Copernicus' heliocentric universe, but it was in geometry that he thought he had found the essential basis of the architecture of the universe itself. He pictured a universe built upon a sort of invisible skeleton made up of symmetrical figures such as circles, squares, equilateral triangles and regular hexagons.

His next step was to explain why such an architectural scheme had produced the Sun and only six planets instead of, say, fifty or a hundred. He wrote that he had a sudden revelation about this particular problem at the end of his first year of teaching at Grätz, when upon a blackboard in front of several pupils he drew a figure, which represents two circles inscribed respectively inside and outside an equilateral triangle. He then had the very strong feeling that the orbits of the planets must be made to fit exactly inside regular solids, which themselves fitted inside one another.

According to Plato, the regular solids are not infinite in number. There are only five of them: the tetrahedron (a pyramid outlined by four equilateral triangles), the cube, the octahedron (eight equilateral triangles), the dodecahedron (twelve regular pentagons) and the icosahedron (twenty equilateral triangles).

Only five perfect solids – and only six planets! This is why Kepler felt that the spheres upon which the planets move must be drawn inside or outside Plato's regular solids. 'The joy which this discovery gave me,' Kepler wrote later, 'can never be described.' And at once he set to work to draft his first book, *The Mystery of the Universe*, in which he gave this great 'discovery' to the world.

By then Kepler was twenty-five years old. His book was drafted in six months. It was divided into two parts, each of which may be regarded as one version of the essential differences between ancient and modern cosmology. In the first part, he developed the idea which he believed to be fundamental: that the great architect of the universe had built the cosmos with the aid of five perfect solids. Kepler

4·2 Kepler thought that the answer to the architecture of the universe lay in geometry. Originally he pictured a universe whose planetary orbits could be calculated by inserting geometrical shapes inside each other. Kepler was later forced to introduce shapes other than cubes and spheres to account for the known facts. It is interesting that Kepler could believe in such concepts while at the same time holding other, accurate theories.

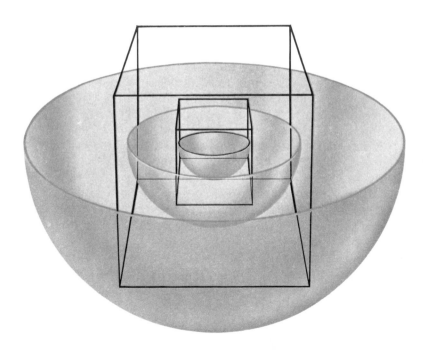

showed how these five solids were arranged one inside the other, with the Sun occupying the centre of the configuration, and how it was possible to construct six spheres which were contained in the overall pattern and which were self-supporting. The orbits of the planets were inscribed on these six spheres.

It is strange that Kepler could have believed in ideas of this sort; moreover, he was forced to contradict himself when he came down to consider details. He knew quite well that the paths of the planets are not perfectly circular, so that they could not possibly be fitted exactly on to the surfaces of regular spheres. Nothing daunted, Kepler gave a definite thickness to the casings of his spheres, so that the planetary orbits could be lodged inside these casings and could be given a certain latitude of movement to either side of the perfectly circular form! The values which he used for the distances of the planets from the Sun were those given in Copernicus' tables, but in some cases it was difficult to make the scale of the model agree with Copernicus' data; accordingly, Kepler calmly declared that Copernicus had been mistaken. He pointed out that Copernicus had not been an observer, so that his figures were unreliable, and could certainly be adjusted so as to agree with the new cosmology. It is quite true that Copernicus made almost no observations, and his figures for the planetary distances were very rough by modern standards; but even so, Kepler's reasoning must be regarded as highly dubious!

A quarter of a century later, in the new edition of his book, Kepler had the grace to retract his slurs upon Copernicus, for whom he had a real admiration. He excused himself on the grounds of youth: 'After all, nobody can take offence at a brat of thirty who makes up his mind to oppose a giant.'

Having described this ideal architecture, Kepler passed on to Part 2 of his book. With no transition at all, he begins the section in a spirit of modern science:

Up to now we have supported our thesis only by arguments of probability. We will now proceed to the astronomical determinations of orbits, and to geometrical considerations. If these do not agree with our theory, then no doubt our previous efforts will have been in vain.

Kepler then set out to calculate something that nobody had ever previously examined: a relationship between the distances of the planets from the Sun, and the velocities at which they move in their orbits. Observational data indicated that the velocity of a planet decreases when the planet is at its greatest distance from the Sun; but why should this be so? Before Kepler, nobody had ever been able to answer that question. Kepler had his own ideas. There should be a force emanating from the Sun, which acted upon the planets and kept them in their orbits; when the planet was farther out, the force would be weaker, and the planet would move more slowly. It even followed that the force should diminish with distance in the same way as the intensity of a luminous source – according to the inverse square of the distance.

Kepler was in the dark here. He tried to link velocity with distance, but at first he did not succeed, and in a chapter of his *Mysteries* he excused himself:

I should have foreseen this from the beginning. However, I did not want to hide from the reader that which will urge me on to new efforts. Let us look forward to the day when the two series of numbers will agree ... I only hope that others will be stimulated to seek the solution for which I have cleared the way.

The New Astronomy

In fact Kepler himself managed to solve the basic problem, but he had to work for another ten years. Then he published his second work, *The New Astronomy*.

Remember that he had never completed his university studies, and in spite of his appointment as Professor of Mathematics he knew very little about mathematics! In particular, he was largely ignorant of trigonometry, which is essential in all astronomical studies. To compensate for this deficiency, he had access to very precise observational data. His great opportunity was offered him by Tycho Brahe, the leading astronomer of the period, who lived at Prague and who had amassed a great number of observations of the Solar System. Tycho had read the *Mysteries*, and was struck by Kepler's intelligence. In 1600, when Kepler was twenty-nine years old, Tycho invited him to work with him in Prague.

Truly this was a case of bringing the wolf into the sheep's fold, because Tycho hoped that Kepler would help him build up his own particular theory, whereas all Kepler wanted was to use Tycho's observations to prop up the idea of regular solids. Tycho himself supported a theory slightly better than that of the ancients, according to which the planets, apart from the Earth, moved round the Sun, while the Earth itself lay at rest in the centre of the world system. The Sun, with its five attendant planets, revolved round the motionless Earth.

Tycho died suddenly only eighteen months after Kepler's arrival. Two days later the Privy Counsellor to the Imperial Prince, Rudolph ii, visited Kepler to nominate him as Imperial Mathematician in succession to Tycho. Kepler accepted, and retained the post until the death of Rudolph ii in 1612. This period in Kepler's life, when he was between thirty and forty-one years old, was the most fruitful in his whole career. It was then that he published his First and Second Laws of Planetary Motion:

1 The planets move round the Sun in ellipses, with the Sun occupying one focus of the ellipse.
2 The radius vector (the imaginary line joining the centre of the planet to the centre of the Sun) sweeps out equal areas in equal times. (This law is often called the Law of Areas.)

Kepler had had to carry out six years' work before reaching these fundamental conclusions. The work had been desperately hard, but it was to be almost as long again before the book containing the results, his *New Astronomy*, was published. The reason for this delay was purely a lack of money to print the book. Scientific progress does not always depend solely upon ideas!

It is impossible to give an accurate estimate of the full effects of these laws upon the astronomical revolution taking place during the period of Kepler's life. At one stroke they disposed of one idea which had been accepted for the previous twenty centuries, and which Copernicus, Tycho Brahe, and – after Kepler – Galileo regarded as inviolate: the dogma of Plato's perfect curves. The first law amounted to a rejection of the idea of circular motion, the second rejected the idea of uniform velocity and all the old theories collapsed in the manner of a house of cards, to be replaced by a more modern picture of the world system.

And yet Kepler did not reject the old dogma as soon as he began his investigations. He took the observational data, and spent some time in trying to reconcile them with the ancient ideas. It was only when he found himself unable to make the

RUDER

BRAHER

BILLER

VLSTANDER

LONGER

RONNOR

ROSENKRANS

TROLLER

AXELLSONNER

LONGER

MARCKEMAN

ROSENSPAR

KABBELER

STORMVASE

GULDENSTEREN

AXELLSONNER

NON HABERI

SED ESSE

EFFIGIES TYCHONIS BRAHE OTTONIDIS DANI
DÑI DE KNVDSTRVP ET ARCIS VRANIENBVRG IN
INSVLA HELLISPONTI DANICI HVENNA FVNDATORIS
INSTRVMENTORVMQ' ASTRONOMICORVM IN EADEM
DISPOSITORM INVENTORIS ET STRVCTORIS
ÆTATIS SVÆ ANNO 40. ANNO DÑI .1586. COMPL.

4·3 Left Tycho Brahe (1546–1601).

4·4 Below Tycho's theory, according to which the planets move round the Sun, while the Sun and Moon revolve round the Earth.

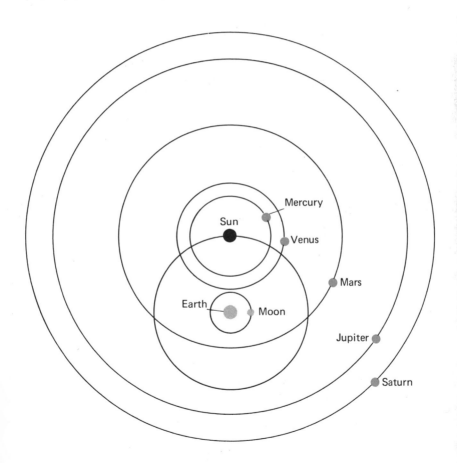

4·5 Kepler's First Law. The planets move round the Sun in ellipses with the Sun occupying one focus of the ellipse. In the diagram the eccentricity of the ellipse is greatly exaggerated.

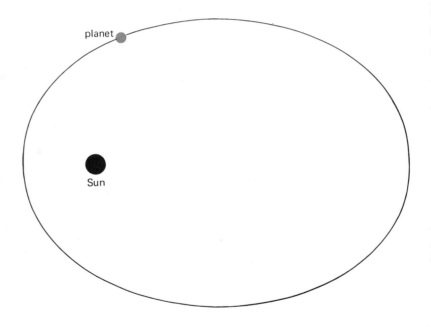

figures agree that the scales were tipped in favour of the modern-type cosmology.

Kepler set out to deduce his laws from studies of the motions of Mars. The choice was a good one, because the orbit of Mars is markedly eccentric. From an observational point of view Mars was a more satisfactory subject than Mercury, whose orbit is even less circular, but which is more difficult to study because of its closeness to the Sun.

Taking Tycho's very accurate measures of the positions of Mars relative to the Sun, Kepler did his best to work out an orbit for the planet. He found that the orbit behaved as though curved, but with the Sun away from the centre of symmetry. For a long time he tried hard to define the curve; at one period he likened it to an oval deformed like an egg, broader at one end than the other. He then considered an ellipse, but for some obscure reason he did not favour it. Yet on 4 July 1603 – that is, in the middle of his six year period of research – he wrote to a friend: 'if the curve could be a perfect ellipse, then I should find all the answers in Archimedes and Apollonius,' and eighteen months later he again wrote to the same friend: 'The truth seems to be found between the oval and the circle, exactly as though the orbit of Mars were a perfect ellipse.' But still he could not make up his mind.

It was the Second Law which was derived first. Wearying of the oval curve idea, Kepler went back to an approximately circular orbit, but he did not put the Sun in the centre; he then tried to explain the varying orbital velocity of Mars in an orbit of this general type. He made long calculations, in which there were numerous mistakes which by pure coincidence happened to cancel each other out. We know about these errors because we possess Kepler's original manu-

4·6 Kepler's Second Law. The planet moves from A to B in the same time that it takes to move from C to D. Therefore the sector A-Sun-B must be equal in area to the sector C-Sun-D.

scripts; there are nine hundred pages of calculations written in a thin, nervous handwriting.

A sudden idea came to him:

> Knowing that there must be an infinite number of points on the orbit, and therefore an infinite number of distances, the idea occurred to me that the sum of these distances is contained in the area of the orbit. I recalled that, in the same way, Archimedes had divided the area of a circle up into an infinite number of triangles.

This was the first step toward the infinitesimal calculus, though only with Newton was it put into mathematical form. Meanwhile, Kepler laboriously added up the distances between the Sun and the path of Mars; he took 180 distances, which he added to each other in order to compare the areas delimited in this manner by the displacement of the radius vector joining the Sun to Mars with the velocity of the planet on its orbit. He re-worked the calculation at least forty times. At last, out of the confusion of all the figures he had accumulated, came the long-awaited answer; it was the Second Law.

Kepler then returned to the form of the path of Mars. He ended up by grouping all the experimental values of the distances of Mars from the Sun as a function of the position of Mars in its orbit, and gave an equation which was nothing more nor less than the equation of an ellipse; but – almost incredibly – he did not realise that he was dealing with the equation of an ellipse! Tired of all this work, he then decided to proceed as if the trajectory were a perfect ellipse, and to see whether, in such a case, the equation of an ellipse would yield positions which were similar to those actually observed. At last he proved that there was indeed agreement. Mars followed a perfectly elliptical orbit, with the Sun occupying

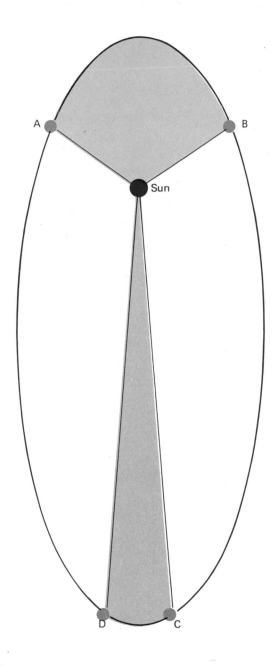

one focus. The same was presumably true for all the other planets – and this was the First Law.

We know the full story of Kepler's difficulties, because he wrote them down with complete frankness. He wrote:

Why chew over my words? The truth of Nature, which I have always sought, comes in stealthily by a side door in a form which makes it acceptable. That is to say, I put aside the original equation and fell back upon ellipses, believing that this was a different hypothesis, though in fact the two were one and the same thing ... I would have had to be mad in order to look for an explicit reason why a planet should prefer to follow an elliptical orbit. Ah! how giddy I have been!

It is a pity that there have not been more researchers who have been honest about the errors which have preceded their discoveries. Frankness would help us to analyse the conditions suitable for scientific creativeness, though it must be said that these conditions are not always very scientific.

Harmony of the World

Even after the announcement of his laws, which are fundamental to modern astronomy, Kepler could not forget his ideas about the architecture of the universe as constructed upon perfect solids. After the publication of *The New Astronomy*, toward the end of 1609, the great astronomer still had twenty-one years to live, but in general he spent this last part of his life in going back to the Pythagorean ideas of harmony, trying to show how the fundamental structure of the universe must depend, as he had believed during his youth, upon an arrangement of invisible perfect solids. In 1618 he finished his book *Harmony of the World*, in which he produced a grand synthesis of geometry, music, astrology, astronomy and epistemology, a little on the lines of what we nowadays call a 'general theory of knowledge'. The work ran to five books. Lost in a welter of purely speculative considerations, we find one vitally important statement. This is the Third Law, which states that:

The squares of the sidereal periods of the planets are proportional to the cubes of the semi-major axes of their orbits.

However, Kepler looked on this law as just a 'detail' in his theory; he was thinking only of his perfect solids when he wrote in his *Harmony of the World*: 'What I have written in my book is for my contemporaries or for posterity. To me, these are equal. My words can wait a hundred years for a reader. God has waited six thousand years for a witness....'

All things considered, Kepler had several reasons for not regarding his laws as the most important part of his work. His contemporaries treated his results with complete indifference, and according to the scientific tradition of the time his perfect solids probably seemed more convincing than his elliptical orbits and varying velocities.

The general reactions of his contemporaries were summed up by an astronomer named Fabricius, with whom Kepler had a long correspondence. After reading *The New Astronomy* and the first two laws, Fabricius wrote to Kepler:

With your ellipse, you abolish the uniformity and circularity of the movements, which seems to me more absurd than I am prepared to admit.... If you would simply retain uniform circular motion, and justify your elliptical orbit by a new epicycle, I would regard your work as being of more value.

Even Galileo, who has always been regarded as a great innovator, always refused to accept Kepler's laws; he preferred to retain the idea of uniform circular motion.

Certain scholars went even further. A contemporary of Kepler's wrote that 'in trying to give material proof of the Copernican hypothesis, Kepler introduces strange speculations which do not come into the domain of Astronomy'.

What can be drawn from all this? After all, it is not science which is the true creator; knowledge must be increased by the work of men – and a pioneer often has few friends to help him or even to approve of him.

In January 1612 Prince Rudolph died. In the same year Kepler lost his wife, his favourite child, and finally his position at Grätz. He was then forty-one years old. However, he was able to obtain a post as Mathematician at Linz, capital of Upper Austria, and he stayed there for fourteen years, leaving only four years before his death.

This fourteen year period was a difficult one for Kepler. First, his post was undoubtedly inferior to that which he had held at Grätz. He knew nobody in Linz; his loneliness was increased by the death of his wife; and his mother was accused of witchcraft and was threatened with burning at the stake, so that he had to make frequent journeys to his birthplace, in Weil der Stadt, in order to defend her. Moreover, he was faced with tribunals which were busily engaged in hunting witches, and which ruthlessly burned even people who had done no more than come under suspicion. (This, let us note, was only a little more than three hundred years ago.) Yet it was under these conditions that Kepler wrote the *Harmony of the World* – which is a paradox in the midst of a social and superstition-riddled climate of this order. It is true, of course, that Kepler had his own way of interpreting the music of the spheres, and wrote: 'The Earth sings Mi-Fa-Mi, from which we must conclude that Misery and Famine dwell in our midst.'

In 1626 the revolt of the Lutheran peasantry resulted in the burning of monasteries and castles, and Linz was besieged. Kepler's house was occupied by soldiers, and the peasants burned down his printing works. Kepler left Linz, and took refuge at a friend's house in Ulm. In 1627 he was called to Prague by Wallenstein, the general who had driven the Danish invaders out of Prussia and who had made it possible for the Imperial Court to return to Prague. Wallenstein believed in astrology, and wanted to draw upon Kepler's knowledge. Thus Kepler, the great astronomer, ended his life in the same way when, as a young mathematician, he had been called to Grätz: he compiled astrological almanacs.

He died after catching cold when riding horseback in the depths of winter. His reason for undertaking a journey on horseback was fantastic: he had set out for Ratisbon to claim twelve thousand florins which the Emperor owed him for arrears of salary. This was 15 November 1630. On the nineteenth he was buried at the cemetery of St Peter, outside the town.

5 A world system
Galileo's single hypothesis

While Kepler was working in Germany, drawing up his new ideas about cosmology, Galileo Galilei in Italy was also considering the universe in a new light. However, the two men went about matters in very different ways, as will be appreciated as soon as we look at their respective personalities.

Even in physique, the two were as different as they could possibly be. Kepler was slight, weakly and a continual sufferer from minor ailments; Galileo was tall and athletic, and remained in good health until nearly the end of his life of seventy-eight years, apart from the fact that at the age of seventy-three he lost his sight.

Above all, the mental outlook of these two giants of science was utterly different. This is shown by the ways in which they wrote their scientific works. Kepler took his readers into his confidence about all the detours and errors which he made during his investigations; he described himself as 'giddy', and he did not hesitate to laugh at himself, as in a letter in which he admitted that he had mistaken a sunspot for the shadow of the planet Mercury seen against the Sun. As soon as new scientific information came to hand, he gave it his immediate attention, and he was quite ready to accept it unless or until it was formally disproved. For instance, Kepler was the only great astronomer who was quick to support Galileo upon the discovery of the four satellites of Jupiter.

Galileo, on the other hand, was always secretive. He said nothing about the ways in which he reached his results, and he even hesitated to make them known, for fear that some other scientist would claim the discovery. On many occasions he used the curious method of announcing a discovery to his friends in the form of a code, which could not be deciphered until Galileo chose to explain the key, but which would establish his claim to priority in the case of any dispute. It was in this way that he sent Kepler the news of his discovery of the satellites of Jupiter; all that Kepler received, in the first instance, was an incomprehensible anagram. He followed the same method when he sent his account of the discovery of the phases of Venus to the Tuscan Ambassador.

Both Kepler and Galileo were distinguished by a great independence of spirit, which marked them out from other scientists of their time. It was this spirit of independence which enabled them to make such tremendous progress in science. In Galileo, however, independence was carried to exaggeration, and caused him to quarrel with most of his contemporaries.

Galileo's usual manner gave the impression of an extreme superiority complex. He seemed to regard all other scientists as simpletons. What made his attitude so dangerous was that he never concealed his opinions in anything he said or wrote, as though he welcomed controversies and legal proceedings. He never gave due credit even to Kepler, whom he held in real esteem. Though Kepler often wrote to him to keep him informed about his results and to ask for his comments, Galileo replied only twice, and generally refused to make any comments at all. In his first reply, written in 1597 after he had received a copy of Kepler's *Mysteries*, Galileo contented himself with saying: 'As yet I have read only the preface to your book', which is a well-known way (still used by some modern scientists) of being deliberately non-committal. Galileo's second letter to Kepler was dated 1610, thirteen years later, and was even less helpful. Kepler had asked for a telescope with which he could see the moons of Jupiter, because, as he stressed he did not doubt Galileo's observ-

ations, and would have liked to have confirmed them personally. Yet Galileo said nothing about the telescope, and Kepler never received it.

Moreover, Galileo never accepted Kepler's laws of planetary motion, and until the end of his days he continued to believe that the planets moved in uniform circular orbits. It is hard to understand why Galileo, who survived Kepler by twelve years, disregarded the laws which may be regarded as the foundations of modern astronomy.

Certainly Galileo did not hide his feelings about the other scientists of his time, notably the Jesuit Fathers. For instance, he wrote the following letter to Cardinal Dini, in answer to attacks launched by those who did not believe in the Copernican System:

The surest way of proving ... the position of Copernicus ... would be to give a whole number of proofs demonstrating that his theory is true and that the opposing theories are false. . . . But how can I do this without simply wasting my time, when those people whom one is trying to convince show themselves incapable of following the simplest and easiest arguments?

And all through Galileo's work we find these strictures upon his opponents, whom he describes as 'mental pygmies', 'stupid idiots', and 'scarcely worthy of the title of human beings'. Those who disagree with him are classed as 'asses', 'elephantine', 'buffoons' and 'poltroons'. Undoubtedly this sort of language made Galileo many enemies, and caused his career to run anything but a smooth course.

Kepler's attitude was quite different. He never concealed his opinions, but he never gave way to anger or irritation, as is shown by the following paragraph from his *New Astronomy*, written in the middle part of his life:

As for the opinions of the saints, I reply that the weight of Authority counts most in theology, but in philosophy it is Reason which matters. Thus St Lactance denied that the Earth is spherical; St Augustine admitted that the Earth is spherical, but denied the Antipodes. Today, the Holy Office admits the smallness of the Earth, but denies its motion. For me, the most sacred thing of all is the Truth; and with all respect to the Church, I have demonstrated according to my philosophy that the Earth is round, inhabited all round to the Antipodes, is of insignificant size, and is travelling rapidly among the other worlds.

Galileo's early life was smooth enough. He was born in Pisa in 1564; his father whose family were minor noblemen, intended him for a career in commerce. However, Galileo showed great zest for study, and at the age of seventeen he entered the University of Pisa, meaning to study medicine. His father was unable to help him, and for some curious reason he was refused a scholarship, in spite of his obvious intellectual gifts. The reason for the refusal was, almost certainly, due to the fact that he was already on bad terms with his colleagues. After he had returned home, Galileo worked upon problems in mechanics, writing his original studies about the pendulum and the hydrostatic balance – indeed, he produced a short memoir on the latter subject, which was quite widely circulated and which came to the notice of Ferdinand de Medici, Duke of Tuscany. The Duke then nominated him as Professor of Mathematics at the University of Pisa, where he had been previously refused a scholarship. Three years later he was nominated for a similar post at the University of Padua; he was then aged twenty-eight, and he held his chair for eighteen years. It was at Padua that he carried out his principal scientific work, the work which helped Newton to draw up a correct model of the world system; this was the study of bodies in motion, known today as the science of dynamics.

The Sidereal Messenger

However, at the age of forty-six Galileo found that dynamics was not sufficient to occupy all his thoughts, and he launched out into a more ambitious subject – that of the world system. In 1610 he published a book, *The Sidereal Messenger*. Thanks to the telescope (which he did not invent, but which he used to good effect) he had discovered the four bright satellites of Jupiter. His fame grew steadily, and he soon took up a new position as Chief Mathematician and Philosopher to the Medicis, in Florence.

In the following year he was triumphantly received in Rome, and was granted audiences by the Pope, Paul v. *The Sidereal Messenger* had dealt with the various discoveries which had overthrown the traditional ideas of the world system; in particular, Galileo had pointed out that the phases of Venus show that Venus must move round the Sun, simply because the phases depend upon the position of the planet relative to the Sun and to the Earth. The Ptolemaic system had been definitely disproved, but there was still a choice between the system of Copernicus and that of Tycho Brahe. Tycho's system, as we have noted, indicated that all the planets except for the Earth move round the Sun, but the Sun itself revolves round an Earth which lies at rest in the centre of the world system.

Galileo had no doubts, and came down emphatically in favour of the Copernican system. As he wrote in the opening section of *The Sidereal Messenger*, it was his firm belief that the Earth must revolve round the Sun.

Contrary to the usual belief, it was not the religious community which first reacted against Galileo's unorthodox ideas. The first attacks came from the lay professors of the universities, who were jealous of Galileo's successes. (Modern universities still contain folk of this kind, who are never ready to accept new ideas, and are always ready to attack any intruder who tries to upset their intellectual house of cards.) The chance came when Galileo published a small tract entitled *Discourse on Floating Bodies*, in which he claimed that bodies floated or sank according to their specific gravity, and not according to their shape; he repeated what had been stated by Archimedes but denied by Aristotle. The university attackers bounded along *en bloc*, and over the next six months they produced four whole books with the object of refuting Galileo's ideas about floating bodies.

Perhaps things would not have gone so far if Galileo had

5·3 Venus in different phases. Galileo pointed out that the phases show that Venus moves round the Sun and could be explained only by assuming that the phases depend on the position of the planet relative to the Sun and Earth. The diagram explains the phases of Venus.

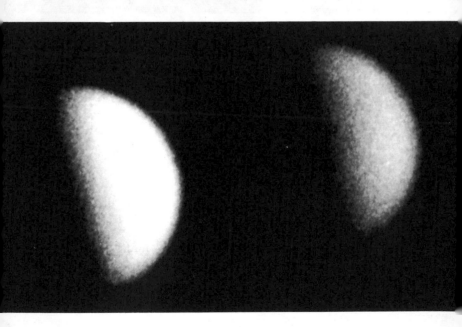

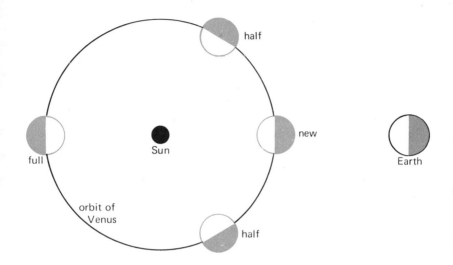

been less fiery, and less alert. Yet his razor-sharp intelligence allowed him not only to demolish his opponents' arguments, but also to make his enemies seem ridiculous. It would take too long to give a full account here of the endless controversies between Galileo and university officials; it was Galileo's bad luck that his enemies were so numerous; as soon as he had disposed of one he found that he had to face another.

Little by little, his adversaries came to understand that the way to overthrow Galileo was not to attack him on subjects such as floating bodies, where the discussion would have to be purely scientific, but to draw him into controversy about the system of the world, where it would be possible to bring the whole might of the Inquisition down upon his head. What had to be done was to lead Galileo into an argument about matters concerning the statements given in the Bible and other sacred writings.

These were the tactics which were put into effect. And, as we know, they succeeded.

Galileo was told that in the course of an evening discussion at the house of the Grand Duchess Christine, several university professors had criticised his cosmological system – which was, to all intents and purposes, the same as that of Copernicus – on the pretext that it contradicted the Sacred Scripture. Immediately Galileo took up arms, and wrote a letter to the Grand Duchess which was passed from hand to hand. In this letter, Galileo began in his usual way, by insulting the 'academic philosophers ... more in love with their doctrines than with Truth ... publishing numerous writings full of vain arguments ... and committing the grave error of adorning their own words with extracts from the Bible which were never meant to be interpreted literally.' After this, Galileo threw himself into a discourse which was much more dangerous. He stated bluntly that the words of the Bible had nothing to do with science, and he added that 'before a physical proposition can be condemned [by the Bible], rigorous proof must be given'. It was this subtle remark which gave the greatest offence to the Fathers of the Church. Not only was Galileo poking fun at the Biblical texts, but he was also saying, in effect, that it should be the Church which ought to give scientific proofs if Galileo were to be faulted. What he did, therefore, was to throw down a challenge which the Church was in no position to accept.

The Fathers of the Convent of St Mark, in Florence, took action, and addressed a combined letter to Cardinal Sfondrati, 'so that the necessary measures should be taken'. Galileo was not actually named, but the famous letter was reproduced. Moreover, there were two misquotes which aggravated the seriousness of the charge. In its quoted form, the letter said that the Scriptures sometimes 'perverted' the facts, though the original letter had said only that the

Scriptures sometimes 'threw a veil' over the facts.

The letter from the Fathers was sent to the Holy Office, but Galileo's popularity was so great that nothing was done about it at the time. And with anyone but Galileo, the whole affair would certainly have ended there. However, knowing that the Fathers of the Convent of St Mark had made an approach to the Holy Office, he decided to present the Pope with his 'physical proof' that the Copernican system was correct. Unfortunately Galileo had no valid proofs to give, and what he wrote is hard to credit, coming as it did from the founder of dynamics. He claimed that the tides are due to the combined motions of the Earth on its axis and the Earth round the Sun. There is no need to stress the weakness of this theory, which can only be described as pathetic. For instance, it could provide for only one tide per day, whereas even the most casual observer could see that Venice has two daily tides.

It is difficult to understand how Galileo could have believed in such a theory – but he did; and, of course, his so-called 'proof', as sent to the Pope, provided ammunition for his enemies. It was only too clear that Galileo had given no proof at all, and that what he had proposed was demonstrably wrong. The Holy Office changed its mind, and put out a categorical statement to the effect that:

1 Copernicus' *Book of Revolutions* must be placed on the Index until corrections to it are made.

2 All similar books teaching the same ideas should also be placed upon the Index.

Note, however, that the Decree did not forbid philosophers and mathematicians from studying, or even teaching, the Copernican system purely as an hypothesis for consideration. What was placed on the Index was the claim that the theory was really true and that the Earth was genuinely in motion, because this was contrary to Holy Scripture, and *because it had not been scientifically demonstrated.*

This last argument is often forgotten, but it was very important indeed. We must never lose sight of the fact that a scientific truth demands a proof, and proof of the Earth's motion round the Sun did not exist in the time of Copernicus, Kepler or Galileo. That the Earth revolves round the Sun is not a self-evident phenomenon; it requires proving. Nowadays there are thousands of aspects of science in which, using simple assumptions, we can put forward new explanations of little-understood phenomena. But science rejects any interpretation based solely upon theoretical or upon experimental demonstrations; it usually insists on both.

Galileo's name did not figure in the Decree, but there could be no doubt that his work was the reason for it. As soon as he knew about the Decree, he wrote boldly to the Secretary of State for Tuscany, stating that 'as can be seen even by the nature of the affair, it is no concern of mine'. However, the Pope instructed Cardinal Bellarmine to 'exhort the said Galileo to abandon his opinions ... if he refuses, he will be imprisoned'.

Actually, matters never reached such a stage. In his report Bellarmine merely noted that Galileo 'had acquiesced'. What had happened was that in order to show his enemies that he had been neither chastened nor humiliated, Galileo had persuaded Bellarmine to give him a certificate to the effect that he had been notified of the Decree, but 'that he had not been judged, and no punishment had been imposed upon him'.

The warning was clear, but Galileo emerged from the affair with honour. The storm was to break sixteen years later.

5·4 Overleaf The frontispiece and title page of Galileo's *Dialogue*, the book which finally led to his famous trial and downfall in 1632–3.

Dialogue of The Two Great World Systems

The Decree was dated 1616. During the next few years Galileo published nothing, but he had not given up the struggle, and as he wrote in a letter to his friend the Tuscan Ambassador he had decided to return to the attack as soon as he was ready to do so. This meant, in fact, finding a definite proof of the Earth's motion round the Sun.

In 1623 Cardinal Barberini was elevated to the Papacy, and became Urban VIII. Barberini had always been friendly toward Galileo, and had advocated a policy of moderation at the time of the Decree of 1616. When Galileo learned of the election, he naturally thought that the time would be ripe to re-affirm his ideas about the movement of the Earth.

He went to Rome, and was given a lengthy audience by Urban, who overwhelmed him with presents and provided a pension for his son, together with letters of introduction to various important people. Galileo saw Urban six times altogether during his stay in Rome, and naturally took the opportunity to speak to him about cosmology – a subject in which the Holy Father was deeply interested. Galileo also spoke about his project of writing a new book about it. We do not know in what terms Urban encouraged Galileo in this project, but that he did encourage him is certain. However, he warned him not to risk bringing the thunderbolts of the Holy Office down upon his head; the Copernican system should therefore be presented purely as an hypothesis. With this reservation, Urban promised Galileo his full support.

Galileo began work in 1626, and by 1630 he had finished the book which was to lead to the celebrated trial: the *Dialogue of the Two Great World Systems*.

After all Urban VIII's promises, it seems extraordinary that

DIALOGO

DI

GALILEO GALILEI LINCEO

MATEMATICO SOPRAORDINARIO

DELLO STVDIO DI PISA.

E Filosofo, e Matematico primario del

SERENISSIMO

GR. DVCA DI TOSCANA.

Doue ne i congressi di quattro giornate si discorre
sopra i due

MASSIMI SISTEMI DEL MONDO
TOLEMAICO, E COPERNICANO;

*Proponendo indeterminatamente le ragioni Filosofiche, e Naturali
tanto per l'vna, quanto per l'altra parte.*

CON PRI VILEGI.

IN FIORENZA, Per Gio: Batista Landini MDCXXXII.

CON LICENZA DE' SVPERIORI.

Galileo should have been led into a situation that ended in his being brought before the Inquisition. To understand how it came about, something must be said about the *Dialogue* itself. It takes the form of a conversation between three people, Salviati the scholar (who represents Galileo); Sagredo, who takes a sensible interest in astronomy and makes a useful foil for Salviati; and Simplicio, who is convinced of the authority of Aristotle, and who is incapable of judging according to the facts.

The three conversationalists discuss the world system for four consecutive days. Gradually the scholar, Salviati, makes the ideas of Aristotle and Ptolemy seem ridiculous; finally he gives a clear demonstration of the truth of the Copernican system. All the time Simplicio is continually abused and made to look stupid. Galileo's sarcastic humour was very much in evidence.

Salviati (that is, Galileo) claimed that all the planets – including the Earth – move round the Sun in uniform circular motion. He showed that this gave a satisfactory explanation of the moons of Jupiter, the phases of Venus, and all the various other discoveries that Galileo had made. It is worth noting that he said nothing about the observations which indicated that the planetary motions are not rigorously circular or rigorously uniform. Kepler's works were completely ignored; in fact, Kepler was not even mentioned.

Incredible though it may seem, Galileo's fourth day in the *Dialogue* led the conversation back to the tides, and once more he presented his definitive proof of the rotation of the Earth. The term 'incredible' is surely appropriate here, for at least two reasons. First, the whole idea was completely unsound, if only because it could explain only one daily tide instead of two – a point which Galileo should certainly have checked. Secondly, the attempt to present a physical 'proof' of the Earth's rotation cut straight across his verbal agreement with Urban VIII. He had assured the Pope that the Copernican system was to be presented as nothing more than an hypothesis; yet now he attempted to give an actual proof, which would change the hypothesis into hard fact.

This was bad enough, but the most serious part of the book came at the end. Simplicio, who had played the role of a clown and who had obstinately refused to understand things which were quite clear to his two companions, was made to withdraw from the discussion in the manner of a whipped dog – declaring that, after all, he could bring himself to accept the Copernican system provided that it were regarded only as *an hypothesis* for consideration. Now, this was exactly what Urban VIII had himself suggested to Galileo, that is, that the Copernican system should be presented in the guise of an hypothesis. Galileo was running an enormous risk. On reading the book, Urban VIII would certainly believe that he was being caricatured; after all, Simplicio repeated the Pope's words almost exactly. In fact, Galileo was making a major tactical error, and was showing yet again that he had a low opinion of other people, judging them incapable of reading between the lines.

The next step was to obtain permission to have the book printed, and Galileo submitted his manuscript to the principal censor in Rome, Father Riccardi. Riccardi saw at once that Galileo had set out to show the physical reality of Copernicus' ideas, and had scarcely bothered to hide his intentions. But Riccardi knew that Galileo was a protégé of Urban's, and so he did not make any prompt decision; he declared the book to be 'beyond his scope' and sent it to another censor, Father Visconti. Realising that the book was potentially explosive,

Visconti made some very minor corrections and sent it back to Riccardi, declaring himself equally incompetent to deal with it. Therefore, Riccardi had to undertake the responsibility himself, but he warned Galileo that the necessary corrections would take some time. As always, Galileo began to argue, and declared that there was need for haste. Eventually a compromise was reached; Riccardi allowed the manuscript to go to the printer, but he reserved the right to correct the work page by page as it came out of the press.

The printing ought to have been carried out in Rome, but at this juncture there was an outbreak of plague, and the Holy City was placed in quarantine. Galileo profited inasmuch as he was able to have the printing done in Florence, and Riccardi was unable to check it, because the quarantine restrictions kept him in Rome. At first Riccardi refused, but at Galileo's insistence he ended by giving way, apart from reserving the right to revise the two most controversial sections of the book – the preface and the conclusion.

Actually, Riccardi controlled only a thin section of the book. Galileo had been too cunning for him. And in February 1632, the *Dialogues* were published.

There was a prompt explosion of rage in the Papal residence and the Holy Office. Urban jumped to the conclusion that he had been mocked, and – rightly or wrongly – he thought that he had been ridiculed by the apparent resemblance between himself and the character of Simplicio in the book.

In August the book was confiscated, and Galileo was summoned to appear in Rome before the tribunal of the Inquisition. Galileo tried to evade the summons, pleading poor health; he managed to hold matters up for a little over a year, but the Inquisition would not let him go so easily, and by 12 April 1633 Galileo was in Rome to face his first interrogation. By now things were very serious indeed. Galileo had gone too far.

This is not the place to describe Galileo's trial, which lasted altogether between August 1632, when Galileo was first ordered to come to Rome, and July 1633, when the verdict was announced. Indeed, to give full details would need a complete book. Neither can there be any attempt here to defend the tribunal, whose conduct is only too clear. But it seems only right to point out one or two facts about the trial which are often overlooked.

First, in spite of the trick that Galileo had played upon him, Urban VIII never lost sight of the fact that he had before him a famous scholar who, despite the present circumstances, had proved his intelligence and even his genius. He also took into account the fact that his prisoner was seventy years old; Galileo was always extremely well treated, he was never put into a cell, and he was certainly never tortured or even seriously threatened with torture. Instead of being confined in a cell during the part of the trial at which he was present (February to June 1633), he was lodged in a five-roomed apartment near the Holy Office. He had his valet with him, and the major-domo attending the Tuscan Ambassador was personally charged with looking after his food and wine.

The *Dialogue* had caused great alarm to members of the Commission, but all historians agree in saying that the final report of the Commission was commendably objective – and it must be remembered that objectivity was not always characteristic of Commissions of the Inquisition. Admittedly, the verdict declared that Galileo must abjure his view that the Earth was really moving, and there was the customary rider about the 'threat of torture', but this was mere convention, and at no time was Galileo in any danger on this score. The expression used in the judgment brought against Galileo was 'territio verbalis', as opposed to 'territio realis' – in the course of which the accused was shown the instruments of torture ... before their eventual use. A few days before the verdict was given the Pope had said to the Tuscan Ambassador: 'After publication of the verdict I will see you again, and we will work out a way to give him as little grief as possible.'

However, it cannot be denied that Galileo was forced to recant before a full Assembly of the Congregation of the Holy Office, and to declare that after all the Earth did not move, so that the Copernican system was nothing more than an hypothesis. We cannot tell what he may have thought; no doubt he was tempted to ridicule his enemies just once more, and he would have been quite ready to re-open the argument. He did propose to the Congregation to rewrite that part of the book which was unacceptable, but the Inquisition contented itself with his public recantation.

Galileo lived for eight more years after sentence had been passed upon him. In 1634, immediately after his trial, he wrote the book which from a purely scientific point of view (not as a history of ideas) is probably his most valuable. This is the *Discourse on the New Sciences*, in which Galileo abandoned cosmology completely and went back to his first vocation, the study of dynamics – a subject which had occupied him for twenty-one years of his long career, as professor first at Pisa and then at Padua.

Tragically, he lost his sight in 1637, but he continued to dictate, and to carry on his work. He died at the age of seventy-eight, held in respect and affection by everyone. His friends wanted to erect a monument to him; Urban refused permission, however, because of the scandal he had caused –

a monument in his honour might encourage others to follow his teaching. Posterity has made amends for this temporary sanction. Nowadays, there is no country in the world in which Galileo is not held in the greatest possible honour. On his broad shoulders, he bore the immense and proud symbol of freedom of thought.

6 An *a priori* world Descartes

With the death of Galileo, we come to the end of the first half of the seventeenth century. And at this point it may be as well to pause in order to take stock of the cosmological knowledge of the period.

The Copernican system had not been adopted without reservation; there were many people who still opposed it, though the only reasons given were that Copernicanism was in open disagreement with Holy Scripture. But gradually the theory of the Earth's motion came to be accepted by European scholars, and also by the general public. The spread of information was due chiefly to Galileo's books, which were unlike most texts of their time, including those of Copernicus and Kepler, in that they were easy to follow. Galileo wrote in language which could be understood by everyone. He introduced characters, such as those of the Dialogue, who discussed everything in a perfectly natural manner; humour was mixed in, and it is easy to laugh at Simplicio in his role of clown. In short, Galileo was not only a scholar but also what would nowadays be called a scientific populariser.

As will be seen, the echoes of Galileo's trial had effects which were contrary to the wishes of the Church. Galileo became something of a legendary figure, typical of those persecuted by the forces of ignorance – and his theory of the Earth's motion became more and more widely accepted.

On the other hand Kepler's three laws, dealing with the elliptical orbits and non-uniform velocities of the planets, were either passed over in total silence (as Galileo had done) or else discussed in a rather lukewarm manner. This distaste for Kepler's ideas was shared by both the ancients and the moderns of the period. The former wanted to preserve the Platonic dogma of uniform circular motion, and so they had no sympathy with Kepler. The latter were equally unenthusiastic, because in Kepler's laws they saw a return to the mystic writings of Aristotle. The force introduced by Kepler to explain the planets' motions seemed to have an occult quality, and this did not fit into current scientific ideas, which Descartes, the next great figure, shared.

Around 1630 it seemed to be generally accepted that all parts of the universe evolve; the old idea of a changing sublunar region and a changeless sky was definitely given up. Tycho Brahe had given a detailed description of the appearance of a new star (now known to have been a supernova); Galileo had shown that Jupiter is the centre of a system of satellites; it had also been found that comets moved in paths far beyond the orbit of the Moon, which was another indication that the whole universe was subject to change.

Discussions were already taking place as to whether the world system is finite or infinite, but great caution was exercised, because here again the problem touched upon the dogma of Holy Scripture. To orthodox churchmen, the world system had been created by God, who lay beyond our world – which must therefore be finite. Earlier, in 1600, Giordano Bruno had been burned as a heretic because he claimed that the world was infinite in both space and time. It is true to say that this particular problem has not been solved even in our own time. It is very difficult to picture a finite universe, because we must also try to picture what lies beyond these limits. On the other hand, it is equally hard to form a picture of an infinite universe which had no beginning.

The science of the first part of the seventeenth century was still coloured by the Platonic idea of a fundamental distinction between the real world, where truth lies, and an apparent world, infected by deceptive human imperfections. Only the real world was important with regard to true knowledge;

6·2 The Mrkos' Comet of 1967. Around
the 1630s such comets were known to be
part of the evidence that the universe
was changing. Descartes was one of the
first philosophers to accept this.

125

there would be found the great principles and fundamental causes. But as the real world was unobservable, all that could be done was to make deductions about it; it was thus that the dogma of uniform circular motion could remain in favour for so many centuries – because despite the observations which so clearly contradicted it, it could be regarded as true in the real world which had been created by God and in which all curves would necessarily be perfect.

It was for this reason, too, that Urban VIII had seriously suggested that Galileo should consider the Copernican system as being an hypothesis only valid for the apparent world. The truth, contained in Holy Scripture, was of course applicable only to the real world.

In short, the laws of nature were relevant only to the real, unobservable world. Observed phenomena were useful in many ways – in navigation, for example – but they did not represent the truth, so that they did not represent true science.

Kepler had started the process of throwing overboard this dualistic conception, because he had showed that the observable world obeyed very precise laws. Galileo, too, had supported the Copernican theory, which depended upon observation; earlier than Newton, he had insisted that truth must be supported entirely by observable evidence. But Kepler had been forgotten, and Galileo had been forced to recant. It was left to Descartes to bring unity back to the scientific vision of the universe.

Rules for the Direction of the Mind

René Descartes was born in France, in 1596. His father had held a post as Counsellor to the Parliament, in Brittany. His mother died a few days after his birth.

The young Descartes quickly showed a taste for studies. Even when he was a mere ten years old he plagued his father with so many questions that he was nicknamed 'my philosopher'. His education was entrusted to the Jesuit Fathers, and at the age of sixteen he began the study of Logic, Physics and Metaphysics. And then, suddenly, he came to some unexpected conclusions, as he mentions in his *Discours de la Méthode*.

As soon as I had completed this course of study, at the end of which time one normally regards oneself as well-educated ... I found myself embarrassed by so many doubts and misgivings that it seemed to me that I had derived no profit from my instruction, except that I had become more and more aware of my ignorance. . . . Above all, I had taken pleasure in mathematical studies because of their certainty and the evidence of their reasoning, but I am astonished that nothing higher has been built upon their foundations which are so closed and so solid ... I will say nothing about philosophy, except that it has been cultivated by the most excellent minds over several centuries, and nevertheless there seems to be no conclusion which is not a matter for dispute. . . . Those things which are not probable I tend to refute as being almost false. With regard to other sciences, which have been built upon the foundations of philosophy, I can judge only that nothing solid can have been constructed upon foundations which are so insecure. . . . This is why I shall completely give up the study of letters as soon as I am old enough to do so, and shall make up my mind to look for another science which I may find in myself or, better, in the world as a whole. I will spend the rest of my youth in travelling ... to gather diverse experiences, to prove myself in the circumstances which fortune presents to me ... I will learn to believe nothing too firmly, and nothing which has been passed on to me by example and custom; and in this way I will free myself little by little from the many errors which blind us and make us less capable of listening to reason.

Descartes' essential idea was to try to obtain a world image which was true, unified and harmonious. The image would have to be the work of one man only, because it was only by human unity that there could be any chance of appreciating the unity of the phenomena of the universe. Therefore, Descartes chose himself as the melting-pot in which to hold this image of the world, and, as he explained later in his *Discours de la Méthode*, he decided 'never to accept anything as true unless I have conclusive evidence that this is so'.

Descartes was thirty years old when he published his *Régles pour la Direction de l'Esprit* (Rules for the Direction of the Mind). This was where he had to begin – since to him, the whole description of the universe was essentially a creation of the mind.

He deliberately decided to spend his time in philosophical reflection. He was fortunate in that his family was well off, so that he had no need to bother about money; and he elected to go and live in Holland, which he judged to be a calmer country than the France of that period, and which he thought to be more progressive scientifically and philosophically. Moreover, Holland had the best publishers in Europe – notably the Elzevir press, to whom Galileo had entrusted the publication of his last works. Freedom of expression was also greater in Flanders and in Holland than anywhere else.

However, Descartes did not sever his links with France. He kept in scientific touch with all European scholars, thanks to his lengthy correspondence with Father Mersenne. Mersenne, who lived in Paris, devoted his life to establishing literary links among the learned men of Europe. He was one of the first to see the necessity of keeping scholars of all disciplines in contact with each other, so that their collaboration would result in a general increase of knowledge. Even today, we have to admit that this lesson has been very imperfectly learned.

The Cartesian method

In 1633, when the verdict of Galileo's trial was announced, Descartes was thirty-seven years old. He was just finishing a work entitled *Le Monde*, in which he applied his rules for the direction of the mind to the study of the universe as a whole, dealing, in particular, with gravity and the movements of cosmical bodies. But for reasons which he explained in his letters to Father Mersenne, Descartes delayed publishing his book, which did not appear until twelve years after his death. The cause was, of course, the sentence passed upon Galileo.

In July 1633 he wrote to Mersenne:

I was intending to send you my *Monde* ... but I must tell you now that recently, in both Leiden and Amsterdam, I have been questioned as to whether the world system of Galileo is included in the book. I hear that Galileo's own book has been printed, but that all the copies have been burned in Rome, and that he has been condemned to some punishment. This seems to me so astonishing that I am half resolved to burn all my papers, or at least allow them to be seen by nobody ... undoubtedly I must try to establish the movement of the Earth ... and I confess that if this is false, then all the foundations of my philosophy are false also.

Naturally, Descartes was not disposed to regard his ideas as faulty merely because the Church had said so, but he was very anxious to keep out of trouble, so that he could continue his work peacefully. On 20 January 1634, he wrote to Mersenne:

I well know that the decision of the Inquisition [against Galileo] should not necessarily be considered as the expression of truth, but

... I am not perhaps so enamoured of my own thoughts ... and I have the desire to live in peace to continue my work, taking for my motto 'He lives well who lives hidden'.... In being so delivered from fear, I shall not be wasting my time ... however, I have not lost hope that my writings will be accepted in the same way as the Antipodes, which were similarly condemned in former times, so that in the end my *Monde* will see daylight.

The *Monde* did indeed appear, but not until thirty years later – in 1662, twelve years after the death of its author. It was in this book, the *Monde*, that Descartes outlined his cosmology. It is striking to see the extent to which Descartes was a forerunner in the manner of summing up the problem, much as Einstein was able to do almost three centuries later. Unfortunately the methods of the time – observational as well as mathematical – were too crude to allow Descartes to give a correct solution of the cosmological problem, however well he set it out.

At this stage it will be helpful to give some details about the Cartesian method of presenting a problem. When this has been done, we can sum up some of his cosmological results.

1 *The axiomatic principle* The first point which makes Descartes a modern is his use of what we nowadays call the axiomatic method. He wanted to build his knowledge of the world upon reason. He knew that this world is very complex; he also knew that we must disregard our own feelings, because these feelings tend to distort the phenomena. In his first Meditation, we read:

All that I have tried to understand up to the present time has been affected by my senses; now I know that these senses are deceivers, and it is prudent to be distrustful after one has been deceived once.

Above all, Descartes prefers to trust his reason rather than his feelings. As a basis he gives a number of axioms, the evidence for which appears to be so strong that there is no fear of being deceived by one's feelings. His famous 'I think, therefore I am' is one of these axioms, and was indeed the most fundamental of all. He then used these axioms as the bases of the rules of reason that he had himself defined; in other words, he made use of logic. He was thus led on to conclusions which were true theorems – inasmuch as they were as true as the axioms upon which they were based.

Descartes was prudent, and – logically – he did not pretend that the world which he described in this way was identical with our real, physical world. He called it 'a new world', a world which could exist only if the axioms upon which it was founded were true in the absolute sense. But Descartes did not tire of drawing up relationships between his ideal new world and the real world. In his heart, he was convinced that his axioms were leading him on toward a description of reality, or at least a good approximation of it. This shows that although Descartes is often regarded as the father of idealism, he was also very much of a realist.

This axiomatic method has, as everyone knows, become the essential basis for interpreting the phenomena of modern physics. If, for instance, we want to study phenomena which are on a very small scale (such as elementary particles) or on a very large scale (such as the universe as a whole), all we can do is to base our theories upon principles deduced from everyday experience and draw up an axiom, according to which we suppose, *a priori*, that things which are beyond our range of observation still obey the same laws as those things which we can study at first hand. Descartes was undoubtedly one of the first to introduce this method into physics, though in geometry it had been used by Euclid and other mathematicians of Ancient Greece.

2 *The geometrisation of physics* Descartes had a very good idea of how to extend his axiomatic principles with regard to the geometry of physics. This was because – once again anticipating Albert Einstein – he believed in what might be called a 'geometrisation' of physics. The idea was not new. Long before, Pythagoras and Plato had taught that 'All Nature and all the heavens are in the symbols of geometry'. But Descartes was not concerned with a formula dictated by intuition, or even wishful thinking; in the seventeenth century, any formula had to be based upon observable fact. Kepler had begun to demonstrate a nature which was 'written in mathematical language', and this formula crops up again in 1623 with Galileo's *Saggiatore*. But it was left to Descartes to sum up the whole principle when he stated that the world can be described above all 'upon the foundations provided by mathematics, which are so firm and so solid'. In this extreme view Descartes was not followed by those of succeeding generations, who generally preferred a mathematical physics which depended upon the observable, and tended to distrust all *a priori* conclusions. However, the idea had taken root, and was ready to be revived at a suitable moment. It was Einstein who finally echoed Descartes with his affirmation, put into practice in all his scientific work: 'A theory can be verified by experiment, but no experiment can ever lead to the creation of a theory.'

To Descartes, the world was composed of an unique substance which filled the whole universe. To Einstein, the world is made up of the unique substance which is called spacetime. To Descartes, all phenomena can be described by 'figures and movements'; to Einstein, the phenomena of space-time are nothing more than figures (curves) and movements (directions) of space-time. The similarity in thought between these two brilliant men could hardly be more striking.

3 *The principles of conservation* In a third way, too, Descartes was far in advance of his time. In choosing the fundamental principles upon which to base his axiomatic reasoning, he became the first to indicate the importance of the principles of conservation.

Qualitatively – and as was stated later by Lavoisier, with regard to chemistry – 'nothing is created, nothing can be destroyed'. There is a wide gap between this simple statement and the discovery that the principle holds good in physics (and in all other disciplines), where we have to deal with transformations taking place on a vast scale, and where we can ourselves interfere with the processes of nature. But in this ever-changing nature it is very valuable to be able to base our reasoning upon solid foundations, which can then be used as a framework in the description of the transformations themselves.

Descartes soon saw that principles of conservation could be put to good use, and he was the first to propose the conservation of the quantity of motion; but he did so clumsily, because he thought, wrongly, that the conservation was independent of direction. However, he was certainly on the right track – and on reading what he wrote, it is clear that he was misled mainly because, at the same time, he suspected the principle of the conservation of energy, which depends solely upon the amount of the energy under consideration.

Like Descartes, Einstein's axiomatic principles were centred upon the principles of conservation; General

Relativity is essentially founded upon them. Riemannian geometry, which Einstein used to support his theory of gravitation, depends on the invariability of length in space-time, another quantity which remains constant in a continually changing world (more particularly, of course, a world which is changing according to the chosen frame of reference). And all this shows that there is a great deal of common ground between the ideas of Descartes and those of the twentieth century.

A Cartesian cosmology

It is rather surprising, then, to find that after he had succeeded in setting out the problem in a way which was so advanced by the standards of his time, Descartes failed to follow up his groundwork. Instead, he branched off into ideas which, when all is said and done, contributed very little to the progress of knowledge.

In my own opinion, the value of Descartes' work has been badly underestimated. To me, the reason seems to be that we are trying to translate the language used by Descartes to express his results as though the words of this language have the same meaning as those of the language of today. If this is correct, then the dismissal of Cartesian cosmology seems to be essentially due to semantics.

A few examples will show what is meant. As a start, Descartes denied the existence of empty space. According to his axioms, this was a perfectly logical conclusion; he had postulated an unique substance which was material, and which pervaded the whole universe. To Descartes, what we normally call emptiness – that is to say, the absence of observable material – is simply a part of the universe which contains nothing apart from the all pervading material. It follows that he could not accept that truly empty space could exist.

In fact, this view is also that of modern times. It is the view given by General Relativity; to Einstein, too, space-time could never be confused with the absence of everything, and space is of the same nature as matter. Einstein himself, in *How I See the World*, wrote unambiguously on the subject: 'Space is real, in the same way as material objects,' and again: 'Space which is empty of matter does not exist.' To suggest that Descartes was reverting to the old metaphysical ideas when he denied the existence of empty space is quite wrong, and, moreover, quite unjust. According to both his axioms and his logic, he had every reason for his beliefs, and they contained the germ of an idea which has been carried forward into our own time.

It may be asked why no criticism has yet been made, in the present chapter, of Descartes' opinion that the planets are hemmed in by 'vortices which agitated the heavens'. Certainly there is every reason to ask why he persevered with this idea, since he was fully aware of the scientific knowledge of the time – in particular, of Kepler's laws, in which the concept of force was defined with admirable clarity.

The reason is that Descartes was unable to accept the notion of force in a scientific description of phenomena. In effect, force is action at a distance – propagated across the void, without any intervening medium; this seems to give it an occult, mediaeval property which Descartes was not disposed to accept. It seemed too mysterious, and in any case there was no evidence for it. Descartes therefore concluded that a planet moves round the Sun because of an interaction between itself and the medium in which it is contained. This seemed, to him, to be more plausible than postulating an unknown force which came from the distant Sun and was transmitted across empty space.

His reasoning was logical enough. Kepler before him, and Newton afterwards, were similarly uneasy about the concept of action at a distance. Kepler discussed it in a way which showed that after he had himself introduced the idea of force, he gave some consideration to vortices, rather in the manner of Descartes. In Kepler's *New Astronomy* we read:

> This kind of force ... cannot be something that spreads out in space between the source and the moving body, but can be something that the moving body receives from the space which it occupies ... nothing is received except where we find a moving body, such as a planet.

Newton, in his celebrated *Letter to Bentley* dated 25 February 1692, said exactly what he thought about action at a distance: 'It is inconceivable that inanimate matter, without the intervention of any material substance, should be able to act upon another piece of matter without mutual contact.' To him, this was so great an absurdity that he did not believe that any intelligent man could believe it.

Once Descartes had rejected this 'absurd' idea of action at a distance, there was only one logical alternative. He concluded that the planets must move along orbits controlled by the action of the space surrounding them. The particular state of space which could fix a planet in a set orbit was termed a vortex – and here we come back to our semantics; if we use the term vortex in its modern sense, then Descartes was completely wrong. (The theory of vortices is one of the most complex in mechanics.) But when the ideas are put into their seventeenth-century setting, they look very different.

Descartes simply deduced that space had particular properties in the region in which a planet was moving; these properties, when interpreted in the forms of 'figures and movements' – that is, geometrically – led him on to the concept of a vortex.

But to make the position clearer, we must turn back to Einstein's General Relativity, which dealt in detail with all forms of gravitational interactions. Einstein rejected the whole concept of force. Instead, he concluded that the planets move in their orbits according to the geometrical properties of space-time in the region in which the planets lie. To explain how each particle of material in the planet can be affected by an outside medium, Einstein wrote that each particle must be regarded as moving along 'lines of force' in space, known as geodesics. To refuse to see the similarity between Einstein's geodesics and Descartes' vortices is to do Descartes a grave injustice. The principles are not so very different in the long run.

It has been helpful to stress this *a priori* vision of the world, introduced by Descartes, because it marks the start of a new scientific era. Several decades later there came the Newtonian picture, which set out to be perfectly phenomenological – that is, based solely upon observation, and rejecting all *a priori* conclusions. Both methods have been used to build up modern science; they may appear contradictory at first sight, but in fact they are complementary. The progress of knowledge is supported in turn by observation (the senses) and reason (the process of thought).

Descartes preceded Einstein in the methods used in Relativity; Newton preceded the Copenhagen School (Bohr, Heisenberg and their colleagues) with regard to the methods used in modern quantum theory. To build up a sound cosmology we must use not one method or the other, but one method *and* the other. This is the way to arrive at a vision of the world which will be both unified and harmonious.

And with these comments, we are ready to pass on to the work of another great man in the history of ideas: Isaac Newton.

7 A world subject to one law Newton

In our own time we have become extremely good at cataloguing people and putting them into definite classes. For instance, Descartes is thought of as a philosopher, and there is a certain unwillingness to add that he made major contributions to the physical sciences as well. With Newton, the opposite has happened. He is regarded as the perfect example of an objective physicist, who has thrown off the prejudices of metaphysics, which are apt to mislead true science along blind alleys. In fact, Newton is not thought to be a philosopher at all, and he is even honoured for the false idea that he was the last sort of person to mix science with philosophy.

Nothing could be wider of the mark. Newton's genius carried the imprint of what we may call his personal metaphysics – and in fact he was concerned just as much with metaphysics as with physics. It is quite unthinkable to suppose that anyone with his brain could be preoccupied with the essential laws which govern natural phenomena, from the very small to the very great, unless – like Descartes – he was also interested in the tremendous questions which are posed with regard to the spirit of man. These questions were purely metaphysical: the world, God, life, death, knowledge. . . .

It may well be asked whether metaphysics was totally absent from the idea of Newtonian gravitation. To judge this, the best course is to re-read the words of Newton himself concerning gravity. He said that in his view, there was an analogy with the fermentation of bodies:

These particles have not only an 'inert life' on the basis of passive laws, but also they are moved by certain active principles, such as that of gravity and that which causes the fermentation and cohesion of bodies. I consider these principles not as reputedly occult qualities due to the specific forms of things, but as general laws of Nature, for which the things themselves are formed; the observed phenomena demonstrate their truth, though their causes have not yet been found.

Newtonian laws seem to lead to the idea that the world system is comparable with a great clock. We know a great deal about the 'clock's' mechanism, and we know that all initiative is absent. This, at least, is the world image proposed by scientists who carried Newtonian ideas to excess. It was not the view of Newton himself, who believed that his results went some way towards clarifying some of the great problems of metaphyics. Near the end of his life, he wrote that he felt rather like a small boy who played on the beach and was able to examine beautiful shells or pebbles, but who knew that the immense ocean ahead of him was still completely unexplored.

The first indication of the law of gravitation

Isaac Newton was born on Christmas Day 1642, eleven months after the death of Galileo. His parents lived in the beautiful orchard country of Lincolnshire. And it was, of course, Newton who was destined to synthesise, in a fundamental law, the cosmological works of Copernicus, of Galileo, of Kepler and of Descartes.

It is said that when Newton was a young man, he saw an apple fall from a tree, and realised that all bodies – whether small or large – are mutually attracted to each other. This story is probably true, but there is a wide gulf between this intuitive idea and the drawing up of a definite mathematical law of universal gravitation. Newton did not publish the last of his results until twenty years later. Even so, it may well be that the fundamental principles were worked out when he

was in his twenty-second year, during the time when the Great Plague in London had forced him to go back to his Lincolnshire orchards where he could work in peace.

The plague struck London in 1665, when Newton was a student at Cambridge. The University closed temporarily, and the young scientist returned to his family home for an enforced holiday. 'During this year,' he wrote later, 'I began to think that gravitation ought to extend out to the orbit of the Moon, and I compared the force necessary to keep the Moon in its orbit with that of gravity on Earth.'

The idea of force was being discussed at this time, but the idea of action at a distance, with no observable contact between the two bodies concerned, seemed mysterious and even unintelligible. There was, of course, an obvious analogy

7·2 Newton's cottage at Woolsthorpe in Lincolnshire, where much of his most important work was carried out at the time of the Great Plague.

143

with magnetism, which does function in precisely such a way; the poles of a magnet act upon one another at a distance, either attracting or repelling each other. It was tempting to suppose that magnetism might be linked with gravity, and Kepler had gone so far as to hint at something of the sort. Newton, however, did not. He never confused the two kinds of attraction, explaining that in his view gravitation was a specific property of all matter.

Things were made more difficult because at that time, the modern idea of 'mass' had not been proposed; scientists preferred the term 'quantity of matter', which, needless to say, is not the same thing. However, the law of inertia had been formulated, principally by Galileo, and this at least was available to Newton. It stated that if a body is in a state of uniform motion, it will continue in this state indefinitely, provided that no outside force acts upon it. And in the absence of force, the moving body will travel in a straight line. Newton knew this, but he also knew that the planets do not travel in straight lines; they describe curved orbits. Clearly, some force must be acting upon them – but what is the nature of this force?

Newton asked himself this question, and no doubt he drew the obvious analogy of a stone which is being whirled around on the end of a string; anyone who carries out this simple experiment will feel a force on his hand, due to the stone 'holding' the string in its path. The force here is named centrifugal force, and had been widely studied when Newton began his researches. In particular, Christiaan Huygens had described its mathematical characteristics. The force increases with an increase in the quantity of material in movement (the mass).

So far as the Earth is concerned, the quantitative data are

7·3 Newton's inverse square law. If a weight of mass m is swung round in a circle on the end of a string of length d, with a velocity v, the centrifugal force is proportional to mv^2/d.

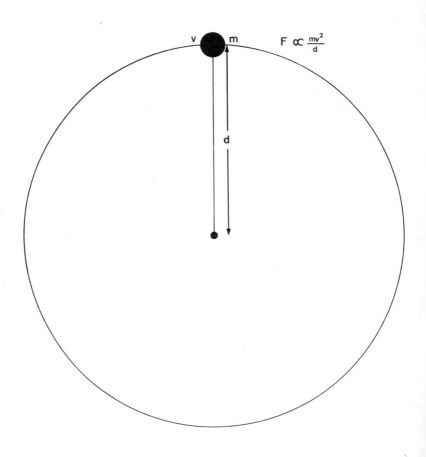

hardly encouraging to the idea of action at a distance. To hold the world in its orbit round the Sun would need a steel cable equal in section to the section of the entire Earth, and stretching all the way between the two globes. It was only sensible to doubt whether the invisible force of gravity could be as strong as so enormous a steel cable.

Newton tackled the problem in a rather different way, starting on the assumption that gravitation is universal for any piece of material, however small it may be. If we calculate the force which the Sun ought to exert on a very small 'packet' of the Earth – a packet weighing one gram, for example – we do not need a steel cable; even the thinnest spider's web would be ample. Therefore, the overall phenomenon should not be discussed as an action of the Sun upon the entire Earth, but as an action upon each small 'packet' of Earth material.

The next step is to verify the idea of gravitation by taking a known example, and checking the answer. It was already known that an object, when allowed to fall, will drop toward the centre of the Earth, which is about 6,370 kilometres below ground level. It was also known – and had been since the mid-seventeenth century – that the mean distance of the Moon from the Earth is about 384,400 kilometres, so that the Earth-Moon distance is approximately 60·1 times the distance between the centre of the Earth and Newton's apple. Newton calculated that to keep the Moon in its orbit there must be an attractive force 3,640 times smaller, by unit of mass, than that on the Earth's surface. Now, 3,640 is approximately equal to the square of 60·1, and so the force of attraction must decrease inversely as the square of the distance. The same was true of the 'force', that is, the intensity, of light. This had already been pointed out by Kepler, and these results were as significant as they were encouraging.

The next step was to show that this attraction is genuinely universal, and that it increases according to the quantities of material concerned. The simplest variation seemed to depend upon the ratio of the masses present, compared with each other. (Actually Newton was still writing about 'quantities of matter', since the concept of 'mass' still lay in the future.)

Newton had set the stage during his time as a Cambridge undergraduate. On these lines, he had intuitively arrived at the formula which made him famous:

the force of attraction between the two masses is proportional to their masses and inversely proportional to the square of the distance between them.

From intuition to principles

In science, it often happens that there is a long delay between the drawing up of an idea which is clear but intuitive, and the final working out of a law which will carry conviction even to sceptics. In the case of the laws of gravitation, the delay amounted to twenty years. Newton did not publish his laws until 1687, when he produced his masterpiece the *Principia*. Newton was then forty-four years of age. Twenty years seems a very long time indeed, but there were two excellent reasons for it.

The first reason was purely mathematical. Newton supposed that any two masses, whatever they may be, must attract each other – but what about the famous apple on the tree in Lincolnshire? It is attracted by the tree below it, by the neighbouring hills, by the rocks below the Earth's surface, by the waters of the oceans – in fact, by everything. The problem is to add up all these attractions and discover

7·4 The gravitational effects of the Earth upon the Moon, which Newton calculated as evidence for his theory of gravitation. In the absence of the Earth, the Moon would move from M to M1. Due to the gravitational attraction of the Earth, however, the Moon is pulled down to M2. The Moon thus takes a circular path as shown because the centrifugal force due to its movement is balanced by the Earth's gravitational pull.

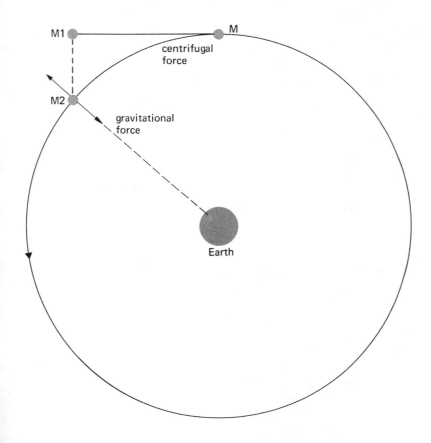

the final result. It means adding up the effects of an infinite number of infinitely small masses lying at different distances. A new mathematical method would have to be worked out, and this is precisely what Newton did. At the age of twenty-five he began to draw up what we now call the infinitesimal calculus, which is so vitally important to physics and to mathematics in general.

Curiously, two other mathematicians – Huygens of Holland and Leibniz of Germany, both of whom were living in Paris at the time – were also making progress along the same lines, but they were working quite independently of Newton, and their notation was different. There was another distinction, too. Leibniz was preoccupied with pure mathematics – in fact, mathematics for its own sake. Newton was, above all, a physicist, and he wanted to use this new mathematical tool to study a problem that was certainly no less gigantic: the problem of universal attraction.

Newtonian gravitation is one of the last historical examples of thought in physics running ahead of thought in mathematics. When Einstein set out to draw up General Relativity, a little more than two centuries after Newton's time, he had the advantage of the mathematical method which had been created half a century earlier by Gauss, Riemann and Lobachevsky: non-Euclidean geometry. Similarly, when Heisenberg was starting his work on the new science of quantum mechanics, he was able to use a suitable mathematical tool – the calculus of matrices – which had also been developed during the mid-nineteenth century by the French mathematician Charles Hermite. Newton, on the other hand, was forced to create the required tools for himself. It was a detail which had to be overcome.

The second reason was even more profound because it was

duc to the prejudices of men of the time. It delayed the publication of Newton's laws, and it also accounted for the fact that the scientific world received the *Principia* with some reserve.

At the beginning of the seventeenth century there had been a sharp, welcome revival of the spirit of scientific inquiry, and the result was a loosening of the grip of Aristotelian physics, which had been maintained for two thousand years. Descartes was one of the most skilful exponents of this new scientific spirit. He was unwilling to introduce into physics the 'occult' idea of action at a distance; science, in his view, should be essentially mechanistic, so that everything could be explained by perfectly discernible and intelligible actions and reactions. It was in this spirit that, as we have seen, Descartes proposed that the Moon and planets owe their motions to vortices which circulate in the heavens and surround all celestial bodies, just as the vortices in a river twirl twigs floating on the surface of the water.

The quality called by Newton gravitation – that is, the property of pieces of matter to attract each other across a distance without any intervening medium – seemed at first to be mysterious, inexplicable, and unintelligible. Moreover, how could an action so intense as this be transmitted across empty space? To most of his contemporaries, Newton was proposing what was, in effect, nothing more than an occult quality of both matter and space. This seemed to be a retrograde step, and a return to the ideas of the Aristotelian school. When Newton's *Principia* was first published, the *Journal des Savants* commented that the book was 'devoid of physical value, because it did not satisfy the conditions required by the intelligibility of the universe'.

There was a long struggle between the supporters of

Cartesian vortices and those who favoured Newtonian attraction before, eventually, the Newtonians triumphed. Yet the victory was not gained because Newton, or any of his successors, proved able to discover the underlying cause of the attraction – and why it could be transmitted across space. It was due to scientific philosophy, which fitted in with Newton's methods, and set out to build a world image which would satisfy observation. This was the theme of scientific method during the following centuries. Newton himself wrote that the cause of gravitation was not important; he observed the movements of the celestial bodies, and he worked out the simplest and best-suited law to explain these movements mathematically, without theorising about their causes. At the end of the second edition of his *Principia*, Newton explained the basis of his scientific philosophy – a philosophy that he regarded as indispensable if knowledge were to increase:

I have explained the celestial phenomena, and those of the sea, by the force of gravitation ... I have not yet deduced the reasons for these properties of gravitation, and I have proposed no hypotheses; because everything that cannot be deduced from the phenomena is an hypothesis, and hypotheses ... should not be contained in experimental philosophy. In this philosophy, *we take the observed facts, and then, by inference, make them general.*

Newton's method is summed up in this last phrase. All the same, Newton's ultimate goal was to explain the causes of the phenomena; his point was that he had been unable to link the causes with the observed facts, and he refused to accept any *a priori* hypothesis. His successors went further. They regarded the idea of cause as purely metaphysical, and therefore unscientific. This attitude has also been much in evidence during the twentieth century; the best example of it was, undoubtedly, the rejection of all causal representation of phenomena in the orthodox version of quantum mechanics. It set up a definite barrier to progress. More will be said about this later.

Faithful to his principles, Newton strongly criticised the Cartesian vortices, which seemed to him to be a typical example of an *a priori* hypothesis not supported by observed facts. It was certain, he wrote in the *Principia*, 'that the planets were not moved by vortices of material. The vortex theory disregarded all observed astronomical phenomena, and raised more difficulties than it explained. To explain the regular, stable movements of planets and comets, it was essential to rid the heavens of all material. A dense fluid would not help in explaining the phenomena of nature; it would merely perturb and hinder the movements of large bodies. Since it made things more difficult instead of easier, there was no proof of its existence, and consequently it ought to be rejected.'

In his biting criticism, Newton – regrettably – distorted what Descartes had thought. When he wrote of vortices of 'material', he meant that Descartes assumed his vortices to be of definite mass, whereas in fact Cartesian vortices were much more nearly similar to Einstein's gravitational 'space-time'. Newton's criticisms would have been more valid if they had been more objective – but in that case they would not have been criticisms at all. He should have made it clear that Descartes had not discussed material vortices surrounding the planets; as it was, he added to the confusion. If he had been more objective, it would have been easier for modern philosophers and physicists to see the bridge across the centuries between Descartes and Einstein.

All this is not, of course, written to try to minimise

Newton's great talent, and the essential place that he occupies in the history of human thought. What I have tried to do is to deal fairly with Descartes. A just parallel between Descartes and Newton, showing that their work was in fact complementary, was given by Fontenelle, at that time Permanent Secretary of the Académie des Sciences, in writing an eulogy of Sir Isaac Newton:

The two great men who were so opposed to each other have, in fact, many points in common. Both were geniuses of the first order, born to dominate other human spirits and to found Empires. Both were excellent geometers, and saw the necessity of introducing geometry into physics. Both founded their physics on a geometry that suited their own principles. But the one (Descartes) was bold enough to outline his ideas, and then interpret Nature accordingly; the other (Newton), more timid or more modest, began by observing the phenomena and then introduced principles to explain them.

The Newtonian world system

Newton's work cleared up a great many outstanding mysteries. Old beliefs could safely be rejected. Comets were no longer puzzling; the Earth did not lie in the centre of the world system; astronomers could look rapturously at the moons of Jupiter and the phases of Venus. All this had been explained by means of this marvellous law of universal attraction, which accounted for the movements of the celestial bodies so precisely.

Newton had achieved a grand synthesis of the knowledge which had been acquired during the previous two thousand years, and he had found the key to some of the greatest secrets of the universe. He had refined Galileo's mechanics, confirmed Kepler's three laws, and invented the calculus to use as a powerful tool in mathematical analysis, so that the laws for small particles could be found and used to extend knowledge of the universe itself.

And yet, as we have seen, Newton's successors went further than the Master himself. Near the end of his life, Newton wrote that he was under no delusions as to the extent of human ignorance. The Newtonian universe was simple only inasmuch as it gave some idea of principles which were very far from simple – that is, the great problems which are essentially metaphysical. These problems have always existed, and always will.

Toward the end of the eighteenth century, before the full benefits of Newton's work had been realised, the great astronomer John Herschel, who founded the science of spectroscopic analysis, confessed that he was somewhat disturbed at how little remained to be discovered by following generations. Did he mean to be taken seriously? Perhaps not; but it is a thought to carry with us as, in our own time, we try to take a more balanced view of the great abyss which separates what we know from what we do not.

Part 2

Relativistic Cosmologies

8 An expanding universe
Two centuries of observations

When Newton put forward his celebrated law of universal gravitation, he gave to the scientific world a mathematical account of the way in which the Moon moves round the Earth and the planets move round the Sun. Yet at the end of the seventeenth century there was no proper appreciation of the dimensions of the universe. All Newton's contemporaries could tell was that the six known planets (Mercury, Venus, the Earth, Mars, Jupiter and Saturn) were in orbit round the Sun; much further away there were the twinkling points known as stars. There was no definite assurance that the Sun was itself a star similar to those visible in the night sky. And so far as the 'beginning' of the universe was concerned, there was general agreement that the only valid evidence was that of the Scriptures: the world had been created six thousand years ago.

The most important problem was to decide whether the stars were fixed in the sky, as had always been supposed, or whether they had individual movements of their own. In 1718 the English astronomer Halley, Newton's colleague, discovered that the positions of three brilliant stars – Sirius, Arcturus and Aldebaran – were sensibly different from those given in the catalogue drawn up by Hipparchus nearly two thousand years before. The most natural explanation was that the stars had moved, and this was Halley's own view; but there was no proof. After all, Hipparchus' measurements might have been faulty. It was only during the second half of the eighteenth century that stellar motions were definitely shown to be real. The man responsible was another great English astronomer, William Herschel.

Herschel was German by birth, and a musician by profession. When still a very young man, he emigrated from Germany to England. He was passionately interested in

8·1 Star-clouds in the Milky Way. The most important problem in the early eighteenth century was to decide if the stars in such clouds were fixed as had always been assumed.

8·2 Below William Herschel (1738–1822).
8·3 Right Herschel's discovery of the planet Uranus in 1781 with a reflector, as shown here, brought him fame. Shortly afterwards he became astronomer to George III and could then afford to devote all his time to his work.

8·4 Uranus and its five satellites.

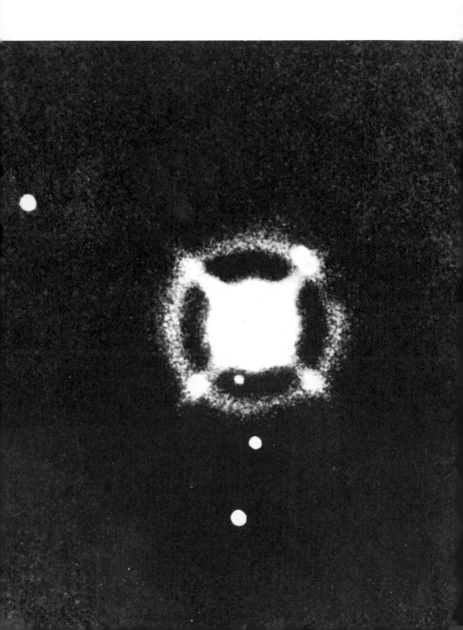

astronomy; as he did not earn enough money as a musician to allow him to buy a refractor, he spent his spare time in making himself a small reflecting telescope, which for an amateur in those days was a remarkable feat. After two hundred failures, he finally succeeded in making a telescope which satisfied him. This was in 1774, when Herschel was thirty-six years old.

At once he set to work on the problem which occupied the attention of most of the astronomers of the period: that of the movements and the distances of the stars. The problem was very difficult. What had to be done was to make very careful measures of the angular shifts of the stars over six-monthly periods, during which interval the Earth would move from one side of its orbit to the other. The diameter of the Earth's orbit was known; if the parallax shifts of the stars could be measured, the distances of the stars could be worked out. In fact, the method was essentially similar to that used by a surveyor who wants to measure the distance of an inaccessible object, such as a mountain top.

As so frequently happens in science, what was found was more important than what was sought. In 1781, after seven years of detailed observations of the stars, Herschel discovered a seventh planet, Uranus, lying further from the Sun than Saturn, which was the outermost planet then known. Immediately he became world-famous. King George III of England and Hanover created the post of King's Astronomer for him, and gave him a salary of £200 per year, which enabled him to spend all his time and energy in studying astronomy; music became a hobby for his moments of leisure. This was clearly beneficial both for astronomical science and for Herschel himself.

He devoted the rest of his life to the study of the stars, notably to their distribution. In 1785 he proposed one of the first reasonably correct models for the star-system; he estimated that the stars were arranged in a flattened system having a form rather like a biscuit. He explained that the Milky Way, that great shining band which stretches across the night sky, is due to the appearance of millions of stars seen in much the same direction. (He thought the number of stars to be between seventy-five million and one hundred million, which is a hundred times less than the number of stars in our Galaxy alone.) Herschel believed that the Sun is a star lying near the centre of the system, but in this he was wrong; again we have to realise that we have no privileged place in the universe. Today, we know that the Sun lies in an arm of the Milky Way, a long way from the centre of the system.

Before the end of the eighteenth century Herschel and the French astronomer Messier catalogued other systems of stars, similar to the Galaxy in which we live. Herschel named them 'island universes' in view of their great distances from each other. However, not all the objects in the catalogues were separate systems; others were the objects known today as globular clusters, which are spherical star-groups lying near the edge of the Milky Way galaxy. There were also objects which were not starry at all, but gaseous nebulae belonging to our own Galaxy.

In 1802 the ever-active Herschel made another important discovery, that of binary stars. A binary is a system made up of two stars, linked gravitationally, and moving round the centre of gravity of the system rather in the manner of the two bells of a dumb-bell. This showed conclusively that Newton's laws applied to the entire universe, not only to the Solar System. It may be added that we now know that about 80 per cent of the stars are members of binary pairs.

8·5 Herschel's view of our Galaxy as shown here was the first to accord basically with the known facts – that the Galaxy is a flattened disc bulging in the middle. It appears in this way because millions of stars are viewed in the same direction.

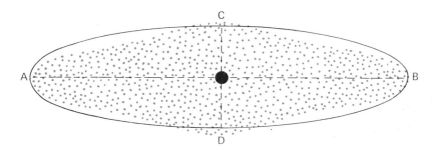

8·6 Herschel believed the Sun to be at the centre of the Galaxy, but the Sun is actually 30,000 light years from the galactic centre.

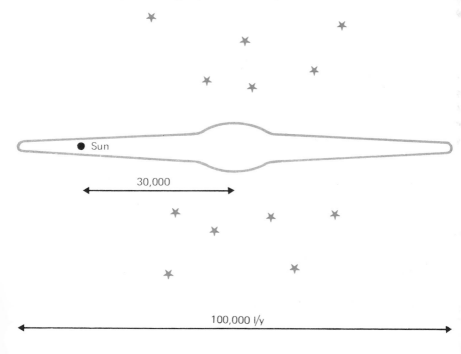

8·7 The binary star Krüger 60, photographed in 1908, 1915 and 1920. Binary stars are two stars linked gravitationally and moving round the centre of gravity of the system. Herschel's discovery of binaries showed that Newton's laws applied to the whole universe, and not only to the Solar System.

1908

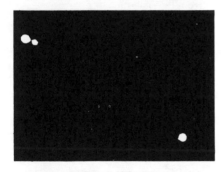

1915

1920

In 1838 the German astronomer Bessel, Director of the Observatory of Königsberg, finally succeeded in making an accurate measurement of the distance of a star, so that the scale of the universe became firmly established. Bessel studied the star 61 Cygni, and found it to lie at eighty thousand milliard kilometres, or five hundred thousand times the distance between the Earth and the Sun – a distance that is 'astronomical' in all senses of the term. At last man began to realise the immense size of the universe, and the sensation of marvellous achievement was not unmixed with awe. During the following century, measurements were made of stars millions of times more remote than 61 Cygni. How much further our range will be extended during the centuries to come remains to be seen; certainly these distances are already so great that the human mind cannot appreciate them.

During the course of the nineteenth century, researches into stellar spectroscopy made considerable progress. In 1814 the German optician Fraunhofer showed that the Sun produces an absorption spectrum; spectroscopy had truly begun, but it was only in the middle of the century that the German physicist Kirchhoff was able to show that the stars are composed of the same chemical elements as those which we know on Earth. Hydrogen proved to be particularly abundant. No longer were the stars regarded as divine; objective science had taken over.

In 1842 the Austrian mathematician Christian Doppler announced the discovery of a principle that was to become of vital importance in astronomy, and is known as the Doppler effect. He showed that light waves received by an observer show a change in frequency depending upon the velocity of the source relative to the observer. The effect is very similar

8·8 Spectra of the Sun (a relatively cold star) and λ Cephei (a hot star). The first spectra showed the stars to consist of the same elements as occurred on Earth, thus removing all thoughts of the 'divineness' of the stars.

λ Cephei

Sun

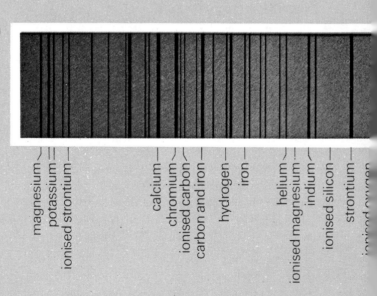

magnesium
potassium
ionised strontium
calcium
chromium
ionised carbon
carbon and iron
hydrogen
iron
helium
ionised magnesium
indium
ionised silicon
strontium
ionised oxygen

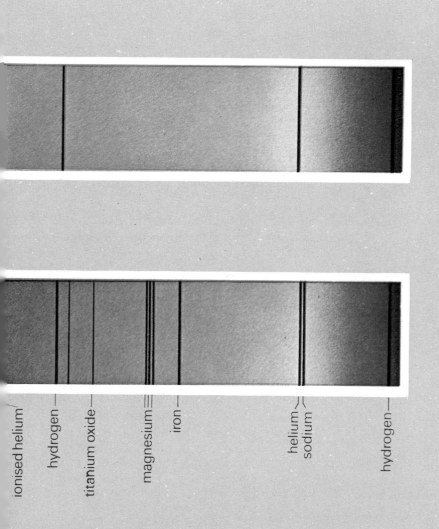

ionised helium

hydrogen

titanium oxide

magnesium

iron

helium
sodium

hydrogen

8·9 Measuring the parallax of a relatively near star. The nearby star is observed at six-monthly intervals, so that it shows a shift against the background of more distant stars. The shift from A1 to A2 gives the star's parallax; since the diameter of the Earth's orbit is known, the distance of the star can be calculated. This method was applied by Bessel to 61 Cygni in 1838.

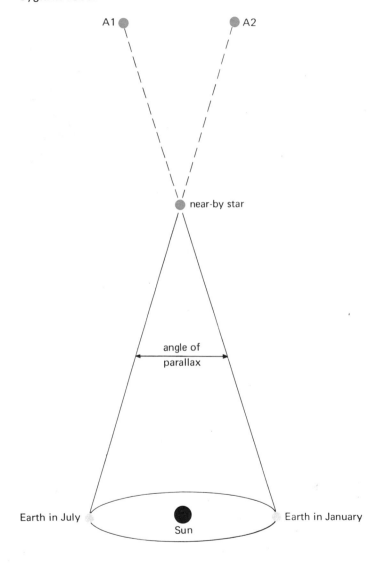

to that which we find with a whistling train; the whistle is high-pitched when the train is approaching, and when the train begins to recede the note of the whistle drops. The light coming from the stars behaves in an analogous manner. If the star is moving towards us, the spectral lines are shifted to the violet or short wave end of the spectrum; if the star is moving away, the shift will be to the red or long wave end of the spectrum. In 1868 the English astronomer Sir William Huggins used the Doppler principle to measure the radial velocity of Sirius, showing that the star is receding at a velocity of fifty kilometres per second. Huggins' investigation was a very delicate one, and was carried through with great skill. Since then, the radial velocities of large numbers of stars have been measured; this particular field of research was regarded as particularly important during the last years of the nineteenth century, and this is still true today.

In 1912 the American astronomer V. M. Slipher, working at the Lowell Observatory, studied the spectrum of the Great Nebula in Andromeda, and announced a velocity of approach of two hundred kilometres per second. However, the Andromeda Nebula is relatively close to us compared with most of the other systems of similar type, and is one of the very few to be approaching us. As early as 1917 Slipher found that of fifteen spiral nebulae investigated, thirteen were receding; the mean velocity of recession was six hundred kilometres per second, which is twenty-five times as great as the mean velocity found for individual stars in our own Galaxy. It seemed that Slipher had discovered a strange phenomenon – that most of the galaxies are racing away from us at great speeds.

The next major step was taken in 1924 by another American astronomer, Hubble, working at the Mount Wilson

172

8·10 The Andromeda Galaxy is the nearest of the really large outer galaxies. It is also one of the few systems to be approaching us.

174

8·11 The light curve of Delta Cephei. There is a relationship between the period of variation and the star's real luminosity. From this it is possible to calculate how far away the star is from the Earth.

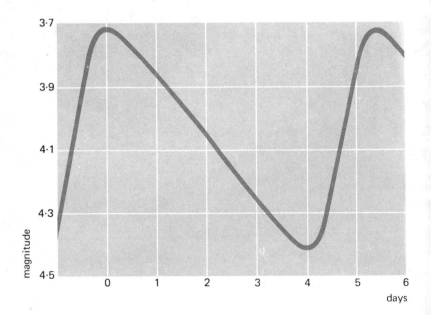

Observatory. In some of the nearest spirals, Hubble detected Cepheid variables, which are highly luminous stars with remarkable characteristics. Many Cepheids had been studied in our Galaxy, and it had been found that there is a simple relationship between the period of variation of the light emitted and the star's real luminosity, since luminosity is a function of the star's distance. This meant that as soon as the period of a Cepheid had been measured, its distance could be found. It was reasonable to assume that the Cepheids in the spiral nebulae would be of the same type as the Cepheids in our Galaxy, and from their apparent faintness Hubble concluded that they must be very remote – at least five times as far away as the most distant globular clusters. In fact, Hubble showed that the spirals were separate systems, well meriting the name of island universes that Herschel had used so long before. Evidently the universe was even larger than had been thought. The distances of the spirals could be measured by the Cepheid method, and this was a great advance, because the spirals are too far away to show measurable parallax shifts; in fact, their distances were unobtainable by the older methods.

Following studies of the velocities of recession of the galaxies obtained by Slipher and later by Humason (Hubble's assistant), and comparing the velocities with the distances of the galaxies obtained by means of the Cepheids, Hubble, in 1930, announced the law which still bears his name. This law states that the speed of recession of a galaxy is linked with its distance from us; the more remote the galaxy, the quicker it is moving away. At that time, the greatest velocity measured was 42,000 km/sec, but by now we have studied galaxies which are moving away at about 150,000 km/sec or 500,000,000 kilometres per hour – over half the velocity of

8·12 Below The Doppler effect. With a body approaching the observer, the wavelength of light is shortened, giving a blue shift. With a receding body, the wavelength is lengthened, and the shift is towards the red.

8·13 Right The Whirlpool Galaxy in Canes Venatici. Hubble's observations of Cepheid variables in such spiral galaxies showed the spiral galaxies to be separate systems and very far away. The universe now appeared to be larger than was previously thought.

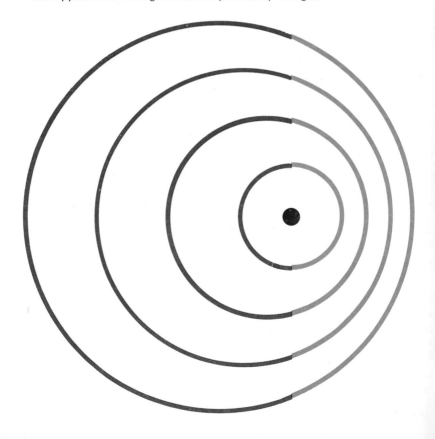

8·14 The universe can be compared with a giant balloon, where marks on the surface, equivalent to galaxies, move apart as the balloon expands. Note that marks already widely spaced show a larger distance increase than closer marks; galaxies behave in a similar fashion.

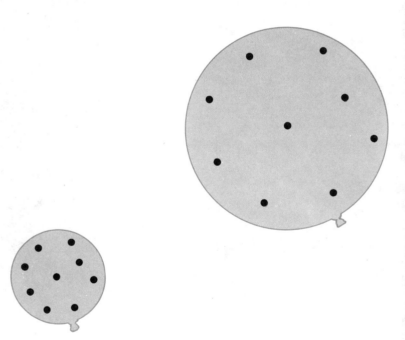

light. Since each galaxy contains thousands of millions of stars similar to the sun, and is of enormous mass, the phenomenon of mutual recession is almost incredible.

Large though the observable universe is, the size is increasing all the time. We may compare the universe with a rubber balloon, with spots painted on the surface of the balloon to represent the galaxies; when the balloon is blown up, each galaxy recedes from each other. Obviously, it is not easy to interpret this remarkable state of affairs. Moreover, what happens at distances so great that if Hubble's Law holds good, the galaxies should be receding at the velocity of light, which – according to relativity principles – is the limiting velocity for any material body? It is here that theorists must come to the help of observers.

Perhaps there is one more fact about the expansion of the universe that is particularly remarkable. As we will see in the following chapters, the expansion was predicted theoretically before being confirmed by observation and experiment.

Over a period of twenty-five centuries man's idea of the world system has undergone many changes, and we have outlined main points of the various cosmologies. However, in spite of these changes there are two points which have remained unaltered since man's first reflections upon the whole complex problem. To describe our universe as a whole, we must select a definite 'substance' which will be the essential framework around which our description is built; moreover, we must find one or several universal laws which will allow us to affirm that whatever may be the fundamental regions of the universe, the fundamental 'motive' applicable to the universal 'substance' will be repeated, and will take on the aspect defined by these laws. In any image of the universe as a whole, we must have these two invariants: a universal 'substance', and one or several universal laws.

To Aristotle, the universal 'substance' was his fifth element, the ether; to him, then, the universal law was that everything should have its proper place, and that every object should return naturally to the place that had been assigned to it.

In the seventeenth century two analogous invariants were worked out, each of which provided the basis for an independent theory. Descartes chose space as the universal 'substance', and constructed his world system upon this alone; he failed to find any laws sufficiently universal for his theory to succeed. Newton, on the other hand, discovered one of these universal laws – gravitation – and was able to understand why the various bodies in the sky move as they do. He also drew up his world system upon this one principle, but he had not found all the answers, because although he had found out the rules governing the motions of material bodies it is not correct to assume that matter makes up the whole of the universe. There is space as well, to say nothing of time.

Newton failed to choose a universal substance which would have been fundamental not only to matter, but also to the space and the time in which the material moved.

In more modern terms, it may be said that to describe the whole universe we must first choose a 'set' with which all the data observable in the universe agrees; this will represent the universal substance. Next we must discover the laws which link all the observed elements of this 'set'; these will represent the universal laws.

Of course, this is a very idealistic approach. In practice, research has to grope its way along; it proceeds by analysis of details, and by inferring a certain number of valid laws from studies of a small number of observed phenomena. As the scope of the investigation widens, principles may come to light that seem to be universal laws; the principle of the conservation of energy is a case in point. It is then the turn of deductive reasoning to play a major role. Starting from the principle that has been found, and using logical arguments, the observed phenomena might be understood, and it becomes possible to predict additional phenomena which have not actually been observed. The principle which is claimed to be universal stands or falls according to this test. If the observations do not confirm all the deductions, without exception, then the principle is not universal. Little by little there emerges a picture of the universe as a whole – in other words, a conception of the unity of the universe.

Probably Albert Einstein's greatest achievement was to realise, and to demonstrate, that a universal substance would have to be re-introduced into science, and that this would have to be the fundamental basis of all observed phenomena. His substance was space-time, which was the modern equivalent of Aristotle's ether or Descartes' extended space. To

model his world system according to the substance he had introduced, Einstein thought in terms of universal principles just as clearly as his predecessors had done, and Newton's law of gravitation served him as a model upon which his ideas could be focused. The result was that incredible edifice of human thought which we call General Relativity.

Gravitation according to Einstein

Unlike Newton, Einstein certainly did not begin by looking at phenomena so that he could subsequently give a description of them. He began by considering the way in which man could write down the laws of nature – *all* the laws, and not merely those concerned with some specific subject, such as electromagnetism or gravitation.

His essential idea, which always remained at the basis of his thinking, was that if laws of nature exist, there must necessarily exist a way of recording them so that they remain valid whatever may be the movement of the observer. In fact, this may be said to be the relativistic method of writing the laws. Einstein held that if the laws varied from one observer to another, there could be no criterion for claiming that the laws were true laws of nature – since all movement is relative. There would then be no true laws at all. However, Einstein believed firmly that true laws must exist, and he was ready to push the consequences of this belief as far as possible.

First let us consider observers who are in motion, but who are not in a state of acceleration relative to each other – for instance, the passengers in a train which is moving at uniform velocity in a straight line, compared with a shepherd tending his flock in a meadow crossed by a railway track. This will lead to a completely new formulation of the notions of space and time, and the measures of space and time will lose their absolute character; we have to deal with a length or a time duration which is dependent upon movement in relation to a frame of reference in which the standards of length and duration have been defined. This is Special Relativity.

Once he had shown how the laws ought to be written so that they would apply to two observers who were moving with uniform velocity with respect to each other, it was natural for Einstein to extend the scope so that the laws could be made to apply to observers who are subject to acceleration with respect to each other. This was the goal of General Relativity.

It is worth noting that, in practice, an observer will nearly always find himself being subject to acceleration, if only because Newton's gravitational forces are mechanically equivalent to an acceleration. In the well-known example of the elevator, Einstein showed that in an accelerated lift an observer could well believe himself to be in a state of rest in a gravitational field; this would explain the phenomena observed inside the lift-cage. Therefore, the phenomena which we find on Earth, and which we must use in our search for laws, are phenomena being seen by accelerated observers. And yet it should be possible to discover, for these laws, a 'handwriting' which will be valid whatever may be the acceleration – in which case they will be true laws of nature.

Einstein, as has been stressed, believed implicitly in the existence of the laws of nature, and so he set out to find a way of defining them so that they would be valid for any acceleration of the frame of reference. It was the resulting formulae which represent, perhaps, his greatest scientific contribution. Acceleration produces effects similar to those

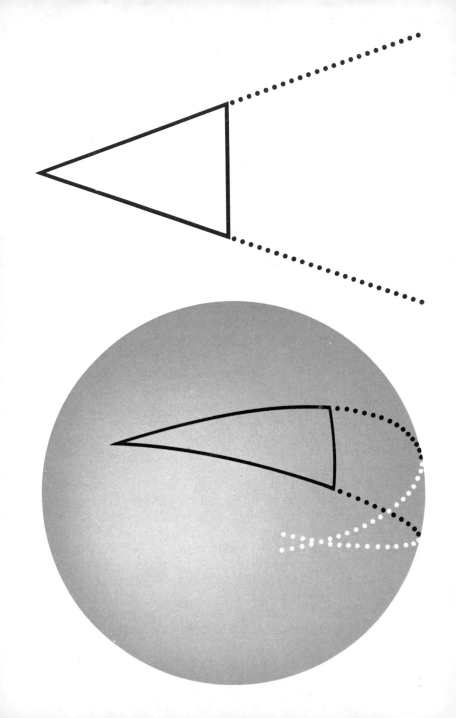

9·2 In the Euclidean system, two lines of a triangle subtended will not join. With curved space, however, the two lines *do* join.

187

of a gravitational field, and so the relationships established by Einstein for General Relativity are directly applicable to the movements of material bodies in a gravitational field. In fact, even though Einstein did not deliberately undertake studies of gravitation, his research led him to provide an expression for gravitational phenomena.

It is important to understand how Einstein's expression differs from that of Newton. As we have noted, some of the researchers who tried to base science solely upon observation were more Newtonian than Newton himself. It is true that Newton declined to put forward any hypotheses about the cause of gravitation, but this was only because he did not consider that he had enough information to guide him, and certainly he was only too anxious to find out the answer to the problem. His attitude here is highly significant. To give a satisfactory explanation, to propose a 'cause' which would not have any hint of an occult quality, it was necessary to go back to the principle which had been proposed by Descartes, and to describe everything in terms of 'figures and movements'. This, surely, is a geometrical description which would have satisfied Newton so far as a description of gravitational phenomena was concerned. His successors, who were not greatly concerned with Descartes' wish for the intelligibility of the laws and merely wanted them to give a correct description of the observed phenomena, went, in fact, much further than Newton. It was Einstein who, in furnishing a geometrical description of gravitation, came finally to Newton's rescue, and – as will be seen – put forward a 'cause' which explains the gravitational phenomena in a pleasingly non-occult way.

At this point it will be helpful to give a brief outline of some of Einstein's essential ideas. So that there may be natural laws valid whatever the gravitational field may be (that is, the acceleration), it must be assumed that space, or more accurately the space-time continuum of Special Relativity, is able to modify its curvature. In regions where there is strong acceleration, space will be more curved than in regions where the acceleration is weak.

There is nothing particularly mysterious about the curvature of space, but we must make up our minds exactly what is meant by the term straight. It is usually assumed that a ray of light will cross space in a straight line until it reaches the observer. Now, a ray of light has the peculiarity that it exactly follows the form of the space. Therefore, space is said to be curved (or, more accurately, non-Euclidean) when it is proved that the light rays crossing it are themselves curved. It may well be asked whether this state of space, of being non-Euclidean, is a hypothesis. For the moment we will be neither more nor less Newtonian than Newton himself, and it is permissible to reject all hypotheses which involve an *a priori* belief that space is Euclidean. Instead, let us find out, by observation, if light rays – which follow the geometry of space – can, in certain regions of space, be curved. If so, we will have transformed a non-Euclidean geometry into space-time. The observation can be made; we can show that light rays passing near the surface of the Sun are curved, so that this region of space has non-Euclidean characteristics. This, in fact, is the celebrated experiment which verified Einstein's idea of non-Euclidean space-time. On its own, Newton's law could not give a quantitative explanation of the results observed; if Newton had been living, he would therefore have accepted General Relativity as an advance upon his own conceptions.

Let it be repeated that there is nothing particularly

9·3 Displacement of a star whose light passes near to the Sun. It should be seen in its real position, but because of the Sun's gravitational pull, the light rays are bent, and the star is seen in its apparent position. Observations of this sort can be made only during a total solar eclipse.

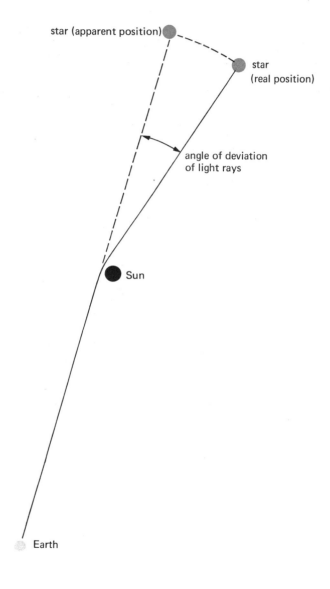

star (apparent position)

star (real position)

angle of deviation of light rays

Sun

Earth

mysterious, unintelligible or even unobservable about this. Once a measure of the behaviour of light-rays in a strong gravitational field has been carried out – and the light-rays passing by the Sun are typical examples – it is possible to visualise the meaning of curved space very clearly and unambiguously.

But Einstein went still further, and was led on to describe the 'cause' of gravitation. To him, matter itself was only a part of space where in a small local region, where the matter was found, there was a very strong curvature. In other words, Einstein discounted the idea of matter being situated in space; there was *only* space – a space where the curvature was very marked at points where there occurred what we call matter. This means that the 'cause' of gravitation is simply the union of the space-curvature between a region where the curvature is at a maximum (that is, at material bodies) and the region of surrounding space. Away from a large mass, such as a star, the curvature of space becomes less, so that a planet moving in this region will to all intents and purposes move according to Newton's laws. The problem of gravitational interaction across the void, with which Newton had been so preoccupied, no longer existed. Space is continuously curved between two material masses acting on each other gravitationally, and the curvature controls the observed movements of the bodies.

Models of the universe

Einstein thus likened the universe to a unique substance, space-time, and his equations gave the universal law furnishing the movements of each point of this substance. Each point was a geodesic of space-time – an idea which, as we have seen, is reminiscent of the Cartesian intuition of trajectories which are the current-lines of vortices.

And yet it is true to say that Einstein did not reach the ideal result of discovering the universal law for every point in space-time. He had discovered the law for gravitational phenomena only, and this was, in effect, a refinement of Newton's law. The more general law, which would have allowed him to describe and predict the movements of all bodies according to their current-lines (geodesics) in space-time, and which would have been valid for all phenomena (notably those of electromagnetism and the atomic nucleus), eluded him. We still do not have what may be termed a unified field theory, but in the future it will doubtless be drawn up, and it will be a complete description of space-time. At least Einstein has shown the way; and in the state where he left it, General Relativity allows us to look at the problem of the universe as a whole more scientifically than has been done before.

It is natural to think that the universe is curved and is given its form by the gravitational field due to all the masses present in it. As we have seen, General Relativity deals with the gravitational problem, and so Einstein's work leads on to the drawing-up of what are called *cosmological models*. Broadly speaking, Einstein's Relativity enables the physicists and astronomers of today to discuss the entire universe in the same way that their predecessors of a few centuries ago discussed the (entire) Earth. Then, opinions were divided between a flat Earth, a spherical Earth, an Earth in the form of a tabernacle, and so on; nowadays, scientists argue about whether the universe is finite or infinite, whether it has or has not been created, whether it is or is not evolving toward a final, predetermined state. It is plain that whenever we consider problems which are at the limits of the knowledge of the moment, the questions asked must relate to metaphysics as well as to physics.

However, to define the overall geometry of the universe, it is essential to draw up a hypothesis. Following Newton's example, it will be wise to reject any *a priori* hypothesis; we must satisfy ourselves that observation not only confirms the hypothesis, but also suggests it. And we must stress that in the universe, we do not occupy a region which, on the average, has any special characteristics. Generally speaking, the mean characteristics of the universe are the same at any point in it, at a given moment of time. This cosmological postulate forms the basis of all models of the universe deduced from Relativity. At least on average, the universe is homogeneous and isotropic everywhere, for any given moment.

Newton might well have asked whether this postulate is a pure hypothesis, or whether it is a consequence of observation. Admittedly, our present-day techniques are not adequate for us to probe to the limits of the universe (if, indeed, there are any limits). On the other hand, optical and radio telescopes can reach out to thousands of millions of light-years, and so far as we can make out the observed phenomena are the same no matter the direction in which they lie. Even the quasars, those immensely powerful, localised sources of energy discovered in recent years, do not indicate any departure from isotropy in the universe; they are found in all directions in the sky.

Still more recently, it has been claimed that the universe as a whole has a temperature of the order of 3° absolute (-270°C.). For various reasons, to be discussed in a later chapter, it is thought that this temperature shows that the universe 'began' in a sort of initial explosion (the Big Bang) a little over 10,000 million years ago; the temperature then was several thousands of millions of degrees, and the universe was both homogeneous and isotropic. Since then, the universe has evolved, but it has retained its homogeneity and its isotropy. To suppose that this is not so would be to make a hypothesis in Newton's sense.

Therefore, we may confidently accept the cosmological postulate. From this, and from the equations of General Relativity, we can begin a scientific examination of the behaviour of the universe *in toto*. At once it is found that the universe has a strange, vitally important characteristic: it is expanding.

10 Models of the expanding universe

When Einstein had worked out the fundamental equations of General Relativity in 1915, he saw their importance in describing the universe as a whole. In effect General Relativity means studying the geometrical form of a unique substance, space-time. This means that the gravitational attraction between two masses can be attributed to a curvature of space-time. This idea can be extended to the scale of the entire universe; it is possible to study what may be called the Relativity of the entire universe, which is assumed to be modelled only upon space-time.

This is what Einstein did in the period following 1917. He adopted the cosmological principle, which states that on average the universe is both homogeneous and isotropic. Starting from the relativistic equations, a spatial radius of the universe, R, can then be calculated. Einstein assumed the spatial radius to be constant (not variable with time), and that the entire universe could be compared with the surface of a sphere of radius R. In this representation, space is reduced to two dimensions only; we are dealing with a surface instead of a volume.

One very remarkable fact that has emerged since Einstein's first cosmological model is that space may be regarded as finite, and yet unbounded. Space is taken as being finite since, in the same way as for the surface of a sphere, it is possible to calculate the volume as a function of the radius R; this volume is not infinite, since R itself is supposed to be finite. On the other hand space must be unbounded. Anyone who could travel 'straight ahead' would never come to a boundary, and would end up by making a complete tour of the universe, coming back eventually to the point from which he had started.

However, Einstein's first model is open to several

criticisms. First, Einstein supposed the universe to be empty of radiation, with a temperature of 0° absolute (as we now know, the temperature is in fact not zero, but 3° absolute). More important is the fact that this model of Einstein's is completely static. It does not change with time, and it does not evolve. This does not agree with observation; for example, we know that stars are born, evolve and finally die, a process which is going on all over the universe.

About the same time (1917), the Dutch astronomer de Sitter proposed another model of a static universe. Instead of being finite, like Einstein's, the de Sitter universe was open – that is, it was of infinite extent in all directions. In effect, the relativistic equations (and the cosmological principle) gave a choice of sign of the curvature of space. If the sign were positive, then space would be spherical and finite, as Einstein had supposed; if the sign were negative, space would be hyperbolic and infinite, as de Sitter had assumed. If the curvature were zero, space would be Euclidean, and we would be back to good, old-fashioned scholastic geometry.

But whether space is finite or infinite, it is clear that both the Einstein and the de Sitter models are static. They take no account of evolution, and therefore they do not correspond to our real, physical universe.

The next important step was taken in 1922 by the Russian mathematician Friedmann. He showed that the static forms of the preceding models were unstable; from the equations of Relativity, it was much more natural to work out a model of the universe whose radius R is not constant, but varies with time. Such a model may be compared with the surface of a sphere whose radius is increasing or decreasing with time. In other words, space would behave in much the same way as the skin of a rubber balloon which is being inflated or deflated, so that the radius is either expanding or shrinking.

However, this model has one feature which at first sight appears somewhat bizarre. The light we receive from a remote star was emitted by the star a long time ago, since when it has been travelling towards us at finite velocity. Therefore, the radius of the universe at the time when the light was emitted was different from its present-day value. Calculations show unmistakably that the difference in the value of the radius of the universe ought to involve a change in the wavelength of the light; broadly speaking, the light would appear either redder or more violet depending upon whether the spatial radius of the universe had increased or decreased since the moment when the light had been emitted from the star. Interpreted as a Doppler effect, this shift ought to lead on to a determination of the star's velocity relative to the observer.

This theoretical forecast seemed so strange in character that up to 1925 few physicists or astronomers would accept it. At that time, it seemed that such a forecast merely showed up the weakness of models of the universe deduced from relativity; all they did, apparently, was to provide an argument for outright rejection of the relativistic models.

However, those who looked closely into the matter saw that to reject the relativistic models would be a very serious matter indeed. It was not only the models of the universe themselves which were at stake; it was the whole theory of General Relativity. The models had been tied to the theory, and to the cosmological principle (homogeneity and iso-tropy), and to cast doubt upon the models would involve casting equal doubt upon Relativity. During the years prior to 1925, however, Relativity had been brilliantly confirmed – for instance by the successful prediction of the bending of

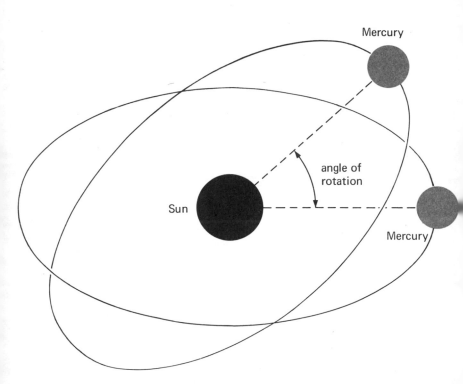

10·1 Two successive positions of Mercury at its perihelion (nearest point to the Sun). Mercury's orbit rotates around the Sun at the same time as Mercury itself rotates around the Sun. The angle of rotation of the orbit during one revolution of Mercury was successfully calculated with the use of General Relativity.

light-rays passing near the Sun and by the observed advance of the perihelion of the planet Mercury. Under these circumstances, it seemed illogical to cast doubt upon the predicted – the expansion or possible contraction of the universe. It, too, might be assumed to correspond to physical reality.

As we have noted, the answer came several years later, with the observation that the galaxies were receding from us. The fact that the galaxies are moving away is shown by the fact that their light shows a red shift – that is, the spectral lines are displaced toward the red or long wave end of the visible spectrum. In terms of Friedmann's model of the universe, this means that the radius of the universe is increasing, so that space is in a state of continuous expansion.

True, there have been various alternative suggestions put forward to explain the red shift. One of these involves a sort of fatigue of the light, which is said to lose energy (and thus increase in wavelength) as it travels through space. However, nothing of the sort has ever been directly observed, so that the idea seems to be a gratuitous hypothesis in the Newtonian sense, and deserves to be rejected. On the contrary, the relativistic theory had predicted the expansion – and this was not its first forecast to be confirmed by observation.

However, we must realise that there are still some aspects of the expansion phenomenon that are far from clear. In particular, there is Hubble's law, which gives the relationship between the distance of a galaxy and its speed of recession. Unlike the principle of the expansion itself, the law is not deduced directly from the relativistic equations. Moreover, it is now fairly certain that the law is more complicated than Hubble originally thought, and it is valid only for distances at which the relative velocity is well below that of the velocity of light. Certainly it is inadmissible to suppose that material structures, such as galaxies, can move in space at velocities greater than the velocity of light (300,000 kilometres per second).

Difficulties therefore remain, notably in connection with Hubble's law; but in principle, the expansion of the universe is accepted today by all astrophysicists, and is not regarded as being in dispute. It remains to be seen what happens at distances where the velocities approach that of light. This is a problem for future research.

11 The continuous creation of matter

One of the most fundamental questions in cosmology – in fact, the question underlying all cosmological studies – is that of the way in which all the matter in the immense universe has been created. Faced with this problem, physicists and astronomers of today find themselves almost as much at a loss as were their predecessors of the previous twenty-five centuries. It is a little like the age-old question: What came first – the hen or the egg?

When my young son had asked me the same question more than ten times, because he found each of my answers unsatisfactory and still wanted to know which came first, I managed to lead him off to some other topic! Yet this characteristic and, perhaps, rather tiring curiosity of a child is also found with adults who have enough time to stop and think. We must, after all, ask ourselves who has created this mysterious and gigantic universe into which we have been placed. Modern science can discuss the creation of the Earth, or even the creation of the galaxies which are spread through space; but in the long run, there is more involved than simple transformations of energy into matter or matter into energy. There must certainly have been a beginning to all the things we see in nature, so that at some time in the past they must have been created. The connection between causes and effects seems doomed to founder before a great cause which justifies itself by itself, and which contains the explanation of its own existence without recourse to outside things. Is this what we call God, as is suggested by purely religious arguments? But who is, truly, God? And is it not rather naive to ask: Who, then, created God? Lucretius, in his *De Natura Rerum*, justifiably commented that 'Nothing is ever begotten from nothing by divine power'.

Questions of this sort have been asked since the early days

of mankind and, naturally, I do not pretend to be able to answer them. However, there are problems which used to belong to metaphysics, but during our own century have passed into the realm of physics. Modern science can probe the universe out to thousands of millions of light years, so that we are seeing regions as they used to be thousands of millions of years in the past. We have become accustomed to dealing with an expanding universe which must have begun (and, therefore, must have been created) more than ten thousand million years ago. It is therefore worth trying to shed light upon how modern scientists set about seeking an answer to the question: How did it all begin?

General Relativity

In ancient times, men asked themselves about the form of the Earth. Was it flat, or was it spherical? This was a problem which remained the object of scientific discussions for many centuries.

Nowadays, with our highly advanced instruments and techniques, the outstanding questions deal with matters on a grander scale. Is the universe, in its entirety, made up of hyperbolic or of spherical space? If the former, the universe will be open, stretching out to infinity in all directions; if the latter, then the universe will be closed, so that anyone travelling straight on for a sufficient distance would eventually come back to his starting point.

We are in a position to consider these possible forms of the universe because Einstein's General Relativity has shown that space is not necessarily Euclidean. It can assume different forms, which depend essentially upon the distribution of the material in the space concerned.

However, it is necessary to make the hypothesis of a principle, known as the *cosmological principle*, which we accept as being valid for the whole of the space which makes up the universe. On average, and taken on a sufficiently large scale, space is, at any moment, homogeneous and isotropic. In other words, on an average there will always be the same number of galaxies per unit volume of space – in all parts of the universe, not only in our own region of it. The motions of the galaxies with respect to each other are of such a kind that there are no preferred directions in space.

It may be said that this cosmological principle appears to be well borne out by observation. Our great telescopes, whether optical or radio, show that conditions seem to be identical no matter which direction in the sky the telescopes are pointed. Even the quasars, those enormous sources of energy which still puzzle astronomers today, are distributed equally in all observed directions in space. In fact, all the cosmological models deduced from Relativity are based upon the assumption that the universe is both homogeneous and isotropic.

Yet although the models rest upon a fundamental hypothesis concerning space, the situation is different when we come to consider time. It is admitted that time is infinite in both the past and the future. What we have to find out is whether the models of the universe involve defining any special or particular moments, in which case it might be possible to relate one of these moments to the beginning of the universe. Significantly, it is found that these special moments do indeed appear in the models.

In this book it is not proposed to give details of the very numerous cosmological models which can be drawn from General Relativity. It will suffice to say that, as has already been noted, the models are divided into two distinct classes: one set of models assumes that space is closed and finite, the other that it is open and infinite. Of course, it is a pity that Relativity leads to this plethora of models, because observational data are not adequate for us to decide which model best corresponds to what we actually see. However, it does look as though the recent data about the absolute black-body temperature of 3 degrees Kelvin, applicable to the whole of space, favour the idea of a spherical, closed universe.

If so, our spherical universe must be in a state of pulsation; that is, its spatial dimensions increase and diminish periodically over long periods of time. At the moment the universe is expanding, as we can tell from the recession of the galaxies (discussed in chapter 10). The beginning of the expansion, corresponding to a universe whose spatial dimensions are reduced to a point, can be dated back to about twelve thousand million years ago.

It is tempting to conclude that according to this model, the universe began 12,000 million years ago, but we must be wary of jumping to conclusions. Even if it is true that all the physical processes we can observe today were born at the time of this 'point-universe' twelve thousand million years ago, time itself did not begin at this moment; it extends into the infinite past. We must conclude that there has been a series of expansions and contractions of space – a universal cycle, so that when the universe has shrunk to its minimum size it immediately gives birth to a new universe which will expand. On average, this cycle will be repeated every 24,000 million years.

According to this model, there is never a true creation of matter (or, more broadly, of energy). When the universe is smallest, the density of energy is greatest, but the energy is always rigorously conserved to the scale of the universe in its entirety. It follows, however, that in the course of a cycle the universe passes through a moment when, at the beginning (or end) of a cycle, space is reduced to a point of nil dimensions. At this moment, the density of energy becomes infinite! Cosmologists usually evade this difficulty by assuming that the minimum dimensions of the universe do not go below a value in which the density of energy in space will be comparable to that in an atomic nucleus; but this explanation is not really satisfactory, because it fixes an absolute beginning of time – at the moment of maximum density. We then have to explain how, at this moment, the material which fills up this original space has been created.

Therefore, it is reasonable to sum up the situation by saying that the cosmological models based on Relativity describe the beginning of the universe, but do not explain how the original material came into being.

A steady state universe

In 1948 three astronomers working at Cambridge University – Bondi, Gold and Hoyle – proposed a new cosmological model, partly in an attempt to avoid this basic difficulty. Their model involved the continuous creation of matter. To them, creation was not one unique act which took place at a well-defined moment in the past; they believed it to be a fundamental phenomenon which has always operated, and is still operating today.

It must be stressed that the Cambridge cosmologists were not basically concerned with the somewhat metaphysical question of the beginning of the universe. By 1948 it had become evident that the cosmological models deduced from

Relativity led to a value of about 5,000 million years since the beginning of the expansion. This was certainly much too low; it was unacceptable because the Sun, whose energy-producing processes were reasonably well known, must be at least as old as that. Bondi, Gold and Hoyle proposed to modify General Relativity so as to reconcile the time-scales given by theory and by observation. The modification was, moreover, draconian. It proposed nothing less than the spontaneous appearance of matter in space over the course of time; and the new matter was assumed to appear from nothingness, thereby violating the principle of the conservation of energy. Yet this principle was at the root of Relativity, and was borne out by all observational data.

It is true that, as the three authors noted, the rate of creation should be very low; it would be of the order of the formation of one proton per cubic kilometre of space over a period of five centuries! As the physicist George Gamow pleasingly remarked: 'at this rate, the creator does not seem to be overworking himself'. However, space is extremely large, and it can be calculated that at such a rate about a hundred new stars would be produced in the universe every second. Bearing this in mind, we are entitled to ask how the creator can muster up so much energy!

Be this as it may, there was an important development in 1952, when W. Baade in America showed that the distances of the galaxies had been grossly underestimated, and would have to be practically doubled. This meant that the time-scale drawn from relativistic cosmological models had to be increased, so bringing them into agreement with observation; the new value for the time-lapse since the start of expansion was over 10,000 million years. This might have been taken as the signal to abandon the continuous creation theory.

Nothing of the kind happened, and the theory is still supported today, though the number of its devotees is declining steadily.

Why do some researchers persist with a theory which violates the sacrosanct principle of the conservation of energy? Probably because mankind retains his strong wish to adopt a spiritually satisfying solution to the problem of the beginning of the universe – and in particular, the original creation of the matter which fills the universe.

As we have seen, Relativity is incapable of giving us an explanation of the creation, simply because it is based upon the principle of the conservation of energy, and cannot therefore admit that material can come out of nothingness. The continuous creation theory proposes that on the scale of the universe as a whole, the principle of the conservation of energy is not valid, even though the rate of spontaneous creation is too low to be detected observationally. If asked who created the material, or how it was created, the answer given is that there is a physical law stating that material can appear spontaneously from nowhere. The idea itself is certainly straightforward enough, but one has to think hard about it.

Moreover, the authors of the continuous creation theory were led on to an extension of the cosmological principle. They supposed that the universe, on average, was homogeneous and isotropic not only in space, but also in time. The newly-created material exactly balanced the matter that vanished from the observable universe. If Hubble's law is valid, galaxies become invisible when they have receded to a distance where their velocities are equal to that of light, and so they would disappear from view. To Bondi and his colleagues, then, the universe was in a steady state; there was no evolution. The universe was homogeneous both in space and in time.

Let us be fair, and agree that since the relativistic models have not provided a satisfactory description of the process of creation, the steady-state hypothesis must be regarded as a useful contribution to the study of this tremendous problem. For this reason it has been worth proposing, and it has caused great interest, even though its weakness in certain aspects (such as its disregard of the principle of the conservation of energy) constitutes a fundamental objection to it. It may well be that a satisfactory theory of the evolution of the universe will have to bear the idea of continuous creation in mind; meanwhile, we cannot yet solve the problem of how the universe came into being.

12 The barrier of cosmic distances

In the preceding chapters we have seen how since remote times, man has tried to form a picture of the universe. Thus over the course of the centuries, various cosmologies have been proposed. Since the middle of our own twentieth century, however, we have entered a new era – that of astronautics. For the first time, man can consider the possibility of leaving the Earth. This is bound to have an effect upon cosmological thought. The adventure of space travel will carry our new Columbus toward the skies which seemed so inaccessible only a few generations ago, and which were once thought to be the dwelling-places of the gods.

On the threshold of this new era, there is one question which it seems that cosmologists should try to answer. How does man fit in with the universe which we are studying with powerful modern techniques, and with the models of the universe in its entirety, which are worked out from the most advanced theories of physics and which suggest that the universe may be finite? Also, is cosmology to deal with the advance of pure knowledge, with man always remaining something of a stranger in this universe which is so vast and so long-lived that it is beyond his comprehension? We must try to decide whether this hardly-won knowledge of the universe must remain theoretical in the sense that the further parts of the Cosmos which we can glimpse with our telescopes will always be inaccessible, even on principle. The crucial philosophical question is, basically, whether we can hope that the wider universe will one day be accessible to us, or whether we must resign ourselves to being confined to a tiny part of it (the Solar System). It is to this problem that we shall now turn.

Some people tend to be astonished by the great velocities now achieved by modern spacecraft. Certainly the velocities are high by everyday standards; they are reckoned in thousands of kilometres per hour – 20,000 to 25,000 km/h for the artificial satellites which move round the Earth, nearly 50,000 km/h for a spacecraft which is designed to go to the Moon and back, and so on. Yet when we consider journeys which involve leaving the Solar System altogether, these velocities appear very low.

To give some idea of what is meant, it will be helpful to consider a scale model, so that Proxima Centauri – the nearest star in the sky apart from the Sun – is placed at the real distance of the Moon. It can be calculated that on this scale, an astronaut whose vehicle travels at 100,000 km/h (about the greatest speed possible at the moment) will be a veritable snail, moving at something of the order of one metre per hour. In fact, a snail ambitious enough to want to travel to the Moon will be in much the position of twentieth-century man who considers trying to reach possible habitable planets moving round other stars.

On our scale, Proxima, the nearest star, lies at the real distance of the Moon. To reach a star on the far side of the Galaxy, our snail will have to cover one thousand times the Earth-Moon distance on our scale model; if it wants to travel to another galaxy, even a close one such as the Andromeda system, it must again multiply the distance by a thousand; to reach a remote galaxy, far away across the universe, a further multiplication by a thousand, so that in all he would have to cover one thousand million times the distance between the Earth and the Moon. Undoubtedly the luckless snail would be doomed to a very long journey indeed!

This crude example shows how astronautics has to face up to the problem of the immense distances separating us from worlds where, without much doubt, there are men who are like ourselves, and who have reached the stage of embarking upon space research.

At a velocity of 100,000 km/h, it would take about 10,000 years to reach the nearest star. Let us be optimistic, and suppose that reaction propulsion will eventually allow us to travel at a velocity of 1,000,000 km/h. A trip to the nearest star will then take 1,000 years, bearing in mind that the spaceship would have to slow down as it approached its destination. Periods of this sort are hopelessly beyond any human life-span, and are even less compatible with the maximum length of time which any astronaut may reasonably be expected to stay on his craft. In practice, I do not believe that manned voyages to other planets can be seriously envisaged if they last for more than five years; and even this makes light of the inherent difficulties of breathing, eating and other essentials.

Then does this barrier mean an insurmountable obstacle in the way of the future development of astronautics, even in principle? Let us look at the matter rather more carefully.

As fast as light

It will be best to begin by considering possible ways of breaking through this barrier. For the moment I do not propose to discuss techniques, but simply to look at the problem in principle. Unless there is some fundamental objection, techniques of the future will be adequate for the task; it is simply a matter of being patient.

It must take a long time to reach worlds lying beyond the Solar System, partly because they are so remote, but also because we cannot travel quickly enough. If it were possible

12·1 One of the most important philosophical questions today is whether we shall be able to observe the universe at first hand by space travel. At present we can make only short trips, as between the Earth and Moon, but we may not be so limited in the distant future.

12·2 A composite photograph of the whole Milky Way.

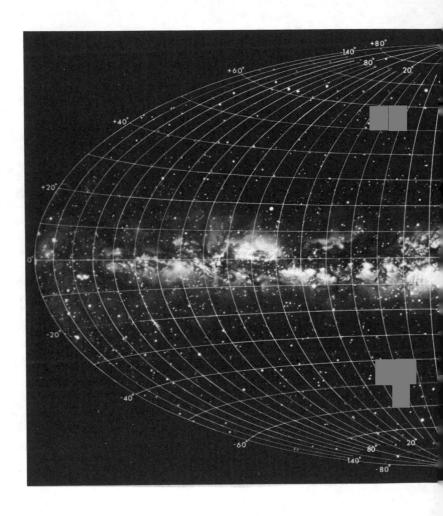

215

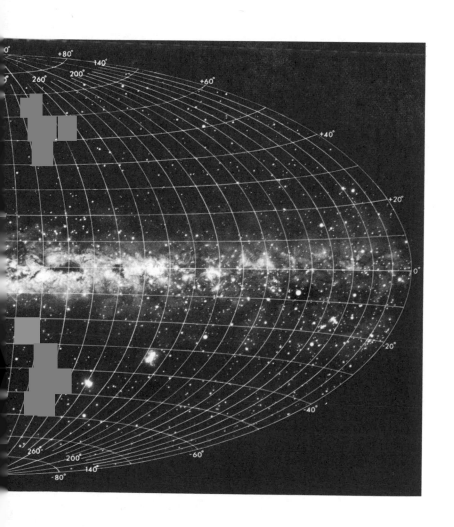

to move sufficiently fast, the time of travel would, of course, be reduced to an acceptable value. In principle, then, it is essential to decide whether a space vehicle can be made to travel at a velocity which is practically infinite.

The answer to the question is given by modern physics, or, more precisely, by Einstein's theory of Relativity. No material mass whatsoever can move, with respect to another mass, at a velocity which is greater than the velocity of light. Admittedly, the velocity of light is very high – of the order of 1,000 million km/h – but it is not infinite. Clearly, then, the velocity of light is the absolute limit of velocity for any spacecraft.

It is tempting to claim that this limit is a pure hypothesis introduced by Einstein as the basis of Relativity theory, and that in the future it may be shown to be wrong. However, to regard the velocity limit as a hypothesis is to misunderstand Einstein's Relativity. The theory is based on the observation that the velocity of light, *in vacuo*, is the same whatever the movement of the frame of reference concerned. Let us repeat that we are dealing with an observation which was known well before Einstein's Relativity was put forward. It is not a debatable fact, and it is not a hypothesis. From this observation, relativistic mechanics can show that the energy of a material body increases together with its velocity, and becomes infinite when the velocity reaches that of light. As no spacecraft can be provided with energy greater than infinite energy, it cannot therefore move faster than the velocity of light. I believe that this assumption is reasonable; any other would be open to strong criticism. To penetrate the barrier of cosmic distance, who would dare to propose an energy greater than the infinite? Even to reach the precise velocity of light would mean giving the spaceship an infinite amount of energy, which would certainly not be an easy matter!

This result is rather discouraging from the viewpoint of possible travel to planets of other stars, simply because even the nearest stars are so far away that a spaceship moving at the velocity of light would take years to reach them. To go from the Earth to the closest of all stars, Proxima Centauri, light takes about four years. Bearing in mind the need to accelerate to the velocity of light (which would take at least a year) and to decelerate before arrival (another year), a space-vehicle journeying to Proxima Centauri would be faced with a travel time of at least six years even if it could reach a velocity almost as great as that of light. And six years is a long time for the crew members to be shut up in their vehicle. Moreover, they could call up by radio after arriving at Proxima Centauri, and send back the information collected, it would be four years before the signals reached Earth. Therefore, no news from them could reach us until ten years after they had set out. I doubt whether astronautics could make much progress under these conditions.

Moreover, it is out of the question to exchange information with planets moving round stars which are further away. In our own Galaxy, most of the stars are thousands of light-years from us, so that thousands of years would elapse between questions and answers, and manned voyages there would take even longer. Beyond the Galaxy, the situation would be even worse, since the delay would amount to millions of years. In short, then, the limitation of the velocity of light constitutes a basic objection to the idea of communicating with worlds lying beyond the Solar System, apart possibly from a few of the nearest stars such as Proxima Centauri.

12·3 The planet Pluto (see arrows) at the boundary of the Solar System.
In order to reach such planets in a reasonable length of time, man
will somehow have to speed up space travel.

Philosophically, it is legitimate to ask why we should be so restricted in principle. In fact, there are certain loopholes to be found, and once again Relativity comes to our help. It may, after all, be possible to shorten the distances if it becomes possible to travel at very great speeds.

Shortening cosmic distances

In theory, and also in practice, we can never exceed the velocity of light, but it may well be that in the future we shall be able to approach it, and to travel at, say, around 1,000 million km/h. Let us picture a spacecraft which is capable of accelerating to this velocity, and that the cosmonauts set out for the Andromeda Galaxy, which lies at about two million light-years (over 10,000,000,000,000 kilometres) from the Earth. During the voyage, several extremely curious phenomena will be observed.

Since the journey will be a long one, the cosmonauts naturally want to make themselves as comfortable as possible during the period of acceleration. On Earth, they are used to the gravitational acceleration g and it is logical to accelerate their ship so that they will experience the same conditions. Therefore, they choose an acceleration of g. They leave Earth, starting from rest; with constant acceleration g, they reach a velocity of 10,000 km/h after five minutes, and are then 1,350 km from the Earth. At the end of an hour their velocity is 130,000 km/h, and their distance from Earth has grown to 200,000 km. After a few minutes more they pass by the orbit of the Moon, at a distance of 380,000 km from their starting point. At the end of their first day of flight, the cosmonauts are going at more than 3,000,000 km/h, and they are not far from the orbit of Mars.

12·4 For space travel over really long distances, as to the distant galaxies shown here, man would have to travel near to the speed of light. At this speed strange anomalies appear in space and time: for example, distance as we know it would disappear.

A week of this fantastic ride will take them to the orbit of Pluto, at the boundary of the Solar System – more than 5,000 million km from home. By then they have worked up to the tremendous speed of 20,000,000 km/h, which would have taken them from the Earth to the Moon in less than two minutes!

If they continue their voyage, keeping up the same acceleration all the time, they will have worked up to more than 650,000 million km/h at the end of eight months – that is, around two thirds of the velocity of light. It is then that they will begin to notice strange anomalies in their navigation. Let us suppose that they are able to measure their distance from their target, the Andromeda Galaxy. They are in the eighth month of their journey, but, contrary to expectation, they find that they have covered a quarter of the total distance; in effect, the Andromeda Galaxy seems as though it had been only one and a half million light-years away from the Earth, though when the spacecraft set out the distance to be covered had been measured as 2,000,000 light-years. The travellers have to decide how, after eight months, they can have covered so much of their total trip.

The situation is so peculiar that our cosmonauts decide to measure their distance from the Earth. Another surprise awaits them. The Earth is unexpectedly close; it is not 500,000 light-years away, nor even eight light-months, nor even at the distance they thought they had covered. It is only a quarter as remote as they had anticipated.

What the cosmonauts have to explain is the remarkable fact that the distance between the Earth and the Andromeda Galaxy seems to have become less than the distance as measured from the Earth's surface before the journey began. The discrepancy is quite large, and amounts to as much as one

quarter of the whole. It is due to the cosmonauts' velocity with respect to the Earth and to the Andromeda Galaxy; because they are moving at two thirds of the velocity of light, the distance is shortened by a factor of one quarter.

This situation was predicted, and perfectly explained, by Einstein's Relativity. *Space is not absolute.* From Earth, we measure the distance of the Andromeda Galaxy as two million light-years; but this distance depends upon the velocity of the observer with respect to the Earth and to the Andromeda Galaxy. If the observer carries out the measurement when he is moving toward the Andromeda Galaxy at two thirds of the velocity of light, the distance is reduced by one quarter. If it were possible to make the measurement with the observer moving toward the Andromeda Galaxy at the exact velocity of light, the distance between the Earth and the Andromeda Galaxy would be found to be zero.

Man as part of the universe

The next step is to try to give a physical explanation of the fact that distances become shortened to an observer travelling at high velocity. Anyone who fails to understand the general principle fails to see that we cannot be dissociated from the universe which surrounds us. It is generally believed that the cosmos is of limited extent, and that we are inside it as independent objects. Actually, we can never be independent of the space which surrounds us; we are merely a 'singularity' of that space.

To give an analogy, suppose that we have a vessel containing water, and that on the surface of the water there is a wave. If this wave is regarded as independent of the water which surrounds it, it will say: 'I am capable of knowing the height of the water in the vessel quite independently of my own height.' But in fact the wave is mistaken – because if it changes its own height, and becomes a large wave, it is bound to lessen the height of the water in the vessel. Therefore, the wave must realise that the height of the water in the vessel depends on the height of the wave itself; this is because the wave is linked with, and is not independent of, the medium which surrounds it. Similarly, man is bound up with space-time, as has been so clearly shown by Einstein. It is wrong to suppose that the distance which separates us from a distant object, such as the Andromeda Galaxy, is absolute. The distance depends upon ourselves, just as the level of the water in the vessel depends upon the wave. If we move toward the Andromeda Galaxy at a high velocity, it will be found that the intervening space has contracted; in other words, the Andromeda Galaxy has approached us.

I deliberately say 'it has approached us', and not 'it seems to have approached us', because this is a perfectly accurate statement. Despite the protestations of the 'absolutists', things are simply what man makes them out to be. Man cannot be dissociated from the universe; the universe is a part of ourselves, though our thought processes sometimes lead us to imagine that we are isolated instead of linked with the cosmos which is all around us.

The universe within our reach

Let us now follow up our description of the cosmonauts' journey toward the Andromeda Galaxy. At the end of a year's flight with acceleration g, they will, at last, be approaching the velocity of light c. Consequently, from the navigators' point of view, distances in the universe are practically nil. At this moment they will want to undertake a manoeuvre similar to that of the first part of their journey; they will want to brake – again, of course, with acceleration g – so that they will have zero velocity when they arrive at their target planet in the Andromeda Galaxy.

This means that keeping acceleration g, they will complete their journey in two years – one spent in working up to c, and the other in slowing down again. Two years is a long time, but so far as a manned space-flight is concerned it is perfectly tolerable.

The important point is that when our cosmonauts have approached the velocity of light as closely as they wish to do, distances will vanish, so that they could reach a galaxy much further away than the Andromeda system without taking any longer on the way. This, of course, is the moment when their velocity is to all intents and purposes equal to c. We have solved the problem of overcoming the barrier of cosmic distances, and the answer is very important philosophically. In principle, at least, we are capable of reaching any part of the universe.

Of course, nothing has been said here about the tremendous technical problems involved in accelerating in the way that we have described. Spacecraft of today can accelerate for only a few minutes, and could certainly not keep up an acceleration of g for a period of a year! But in principle, there is no reason why techniques of the future should not be able to develop spacecraft adequate to the task. So long as the velocity limit c is regarded as the main obstacle, and so long as there is no reason in principle why spacecraft should not be made to accelerate steadily over long periods, it seems clear enough that our scope is virtually unlimited.

Returning from long journeys

The next step is to consider the return journey – for instance, the return of our hypothetical cosmonauts who have been out to a planet in the Andromeda Galaxy. In principle, there would seem to be no difficulties whatsoever; if the spacecraft can accelerate for long periods, the return trip should be no more hazardous than the outward flight. (Again, we can disregard purely technical problems for the moment as we are dealing only with principles.) The cosmonauts make their return journey in the same way as the outward one; they accelerate at g for one year, and then decelerate for one year also at g.

This means that their eventual journey – which may have taken them to the remote depths of the universe – will have taken up only four years; two for the outward flight, and two for the return. Our cosmonauts will come home for a well-earned holiday, with many stories to tell to their children – happy as Ulysses.

A completely erroneous interpretation has been placed upon the return journey, which is supported by many scientists. It has been suggested that on their return to Earth, our cosmonauts will have the surprise of their lives – that of finding that during their voyage to the Andromeda Galaxy and back, the Earth has grown older by four million years!

In my opinion, there is nothing in Relativity to support an idea such as this. The idea comes from a pre-relativistic method of thought; it is believed that the space in the universe is absolute, and that the Einsteinian revolution applies only to the notion of time. This misconception leads on to the idea that travellers moving with respect to each other at high velocities (approaching c) will age differently. To explain how the cosmonauts can arrive at the Andromeda Galaxy in two years, it is supposed that the distance between the Earth and the Andromeda Galaxy always has the absolute value (2,000,000 light-years) that we observe from Earth, but that the cosmonauts age less quickly when approaching velocity c, and do not age at all when c is reached. According to this point of view, they would age only two years on their own time-scale while on the way to the Andromeda Galaxy. When they return home, they would find that the Earthmen, who have not benefited from the slowing down of the ageing process which is associated with high velocity, would have grown older by four million years.

Actually, there is a very simple way of showing that nothing of the sort is possible. In effect, all velocity is relative. When the cosmonauts set out for Andromeda, it can be said that they have a velocity near c in relation to the Earth – but it can equally well be said that it is the Earth which has a velocity of nearly c (in the opposite direction, of course) with respect to the cosmonauts. If any difference in the rate of ageing is involved, it is obvious that in the first case there would be a slowing down of the ageing of the cosmonauts with respect to the Earthmen, but in the second case (which is equivalent to the first) it would be the slowing down of the ageing of the Earthmen with respect to the cosmonauts. Of course, this is a flagrant contradiction, and involves a failure in reasoning.

There have also been attempts to show that there is no symmetry between the cosmonauts and the Earthmen because asymmetrical accelerations are involved. The cosmonauts would accelerate to build up to maximum velocity, and then slow down, and repeat the process for the return trip. However, this could not possibly lead to any differences in the rate of ageing – simply because the cosmonauts have remained under an acceleration of g throughout the voyage, so that from this point of view they have been in exactly the same conditions as the Earthmen.

Above all, it is wrong to suppose that there may be two different types of accelerations, one type produced by gravitation and the other by accelerated motion. Anyone who maintains this does not understand General Relativity; it is only necessary to refer to Einstein's famous 'elevator' example to demonstrate that the two kinds of accelerations are in fact identical.

Yet in spite of Relativity, and in spite of the mass of evidence that makes Relativity Theory the greatest factor in contemporary physics (and even philosophy), there are still some physicists who cling to the past, and persist in thinking in terms of absolute space. For instance, they claim that if the cosmonauts could go to the Andromeda Galaxy and back in only four years, they would be travelling at a velocity greater than that of light – which is a theoretical impossibility. This is a clear confession that the supporters of this idea are still obsessed by absolute space. The cosmonauts have not moved more quickly than light if we assume that the distance between the Earth and the Andromeda Galaxy is not absolute. As measured by the travellers, it is found to be reduced, simply because their frame of reference is moving, with respect to the starting point and the destination, at a velocity almost as great as c. To the cosmonauts, then, the distance is shortened, and is covered in only four years, even though the spacecraft never exceeds the velocity of light.

To make a light-signal cover the outward and return journeys as quickly as our cosmonauts, all that need be done is to make the light-source move in the same way as the cosmonauts in their rocket. The light-photons will then 'see' a shortened distance. The source is first accelerated toward the Andromeda Galaxy until it almost reaches c, so that for this light-source the distance between the Earth and Andromeda is reduced to two light-years, and the light will take only two years to reach Andromeda. The same will apply to the return signal, emitted from a source moving from Andromeda toward the Earth.

What is man?

'What is Man in Nature? A nonentity, with respect to infinity,' wrote Pascal. Certainly it is true to say that taken on his own, man is insignificant with regard to the universe as a whole. But looking at the matter more closely, we must see the direct links which connect us to the whole of the universe, just as a drop of water is bound to the ocean. We can adapt ourselves in space-time beyond the limits that we normally regard as conventional, and we can take advantage of the knowledge of the universe that we have gained.

After all, what is man? Why are we so anxious to regard him as insignificant – something that has developed by sheer chance in a small corner of the universe? It is not conceit to claim that human thought is the basis of everything that surrounds us. Neither is it conceit to say that the human phenomenon is spread through the universe in the same way as the phenomenon of matter – the star phenomenon, for instance. And we can see that as evolution goes on, the network of human thought will lead us from end to end of a cosmos which is completely accessible to man. Pascal knew this quite well, and indeed he wrote:

It is dangerous to overstress man's equality with beasts, without also showing him his greatness. . . . Man is only a reed, the weakest in nature, but he is a thinking reed.

13 Has the universe a destiny?

Man sees in front of him the opening doors of a living cosmology. In the future he will form an idea of the universe which is within his range; there is no insurmountable barrier to prevent him from seeing the whole of the cosmos which surrounds him. And yet now, as formerly, cosmology sets another fundamental question. The universe extends not only in space, but also in time; it stretches back into the past, and forward into the future. We must decide whether we can yet give any truly scientific answers to the problem of the flow of time, on the scale of the whole universe. It was in discussing this problem that St Augustine made his celebrated remark:

What is time? If nobody asks me, I know; if I want to explain it to those who ask me, I no longer know. I want indeed to be able to understand this great mystery.

Unfortunately, the problem of time is as mysterious as ever, despite the Einsteinian revolution. Did the universe ever have a beginning, and what will be its destiny, if it has a destiny at all? Thinkers of past generations have put forward their replies to these questions, but the answers are always enigmatic, vacillating and imprecise. Moreover, all the answers are drawn more from intuition than from reasoning.

The past

When we want to study the past history of the universe, and in particular the vexed question of its origin, we must turn to General Relativity. At least we have something to guide us. Factual observation has confirmed the relativistic data upon two principal points, the expansion of the universe and its absolute temperature.

1 *The beginning of the expansion* If the universe is in a state of continual expansion, as is indicated by the spectral red shifts of the galaxies, it follows that in the remote past the universe was much smaller than it is now. We can even work backwards, so to speak, and calculate the time when the radius of the universe was nil. It is logical to regard this moment as marking the birth of the universe. From the observed data, it can be calculated that the birth occurred from 10,000 to 12,000 million years ago. However, this figure must be taken with a certain amount of reserve, because during the past thirty years – ever since the expansion was discovered – it has been steadily increased. The original estimate was four to five thousand million years, but by now the accepted value is two and a half times greater. (It is also true that in Newton's time, scientists in general believed that the world could not be more than a few thousand years old.)

It may be assumed that around the time of its origin the universe was very small; it may have been comparable with the size of the present-day Sun. Its temperature was presumably several thousands of millions of degrees absolute. In such a furnace, particles of matter could not exist, and everything was in the form of electromagnetic radiation. Here, then, is a case in which Scripture and science agree; light was created in the earliest days of the universe.

Gradually this radiation was changed into matter – that is, into protons, neutrons and electrons, which are the stable elementary particles studied by modern physicists. By gravitational forces, this great cloud of expanding matter broke up into numerous smaller clouds, which were to become the galaxies we know today. Inside these proto-galaxies, gravitational forces continued to produce condensations of material; it was these condensations which were to become stars. The stars began to radiate, because of thermonuclear reactions going on inside them. The reactions are well understood today, and may be said to be essentially similar to the reactions produced in an H bomb.

Around the stars, the scattered material formed new condensations, which became planets. And on certain planets, such as the Earth, evolution progressed from inanimate matter to vegetation, vegetation to animal, animal to man.

Of course, this description of the evolution of the universe is very over-simplified, and many of the points are still highly problematical, but the vital fact is that the universe had presumably a definite moment of creation. At birth, it was homogeneous and isotropic, since it consisted of radiation at a colossal temperature. It follows that on average, evolution proceeded along the same lines in all parts of the universe. Modern cosmology, which takes great care to be suitably scientific, indicates that thousands of millions of Earths similar to our own are scattered all over the cosmos. This means that the human phenomenon applies on the scale of the whole universe, and is not merely an insignificant event developed by chance on a planet which is singled out from the thousands of millions of others. The self-centred view that we are the only men in the universe is regarded nowadays as totally unacceptable, except as an hypothesis which has no observational basis. Apparently it was condemned long ago by Newton, the father of modern scientific thinking.

However, to say that the universe began 12,000 million years ago is not to answer the essential metaphysical question which is always to the fore: Who created the universe? Or, to put it in another way, how could the universe come out of nothingness, since the expansion indicates that it passed through a phase of nil volume at the moment of its birth?

Modern science has given various replies to this question, but all the replies leave a good deal to be desired. For instance, there is the sacrosanct principle of the conservation of energy, which makes it difficult to explain the transition between nothingness and the enormous energy which we can measure in the universe today. On this point, modern science has scarcely improved upon the evasive idea proposed by Anaximander, six centuries before Christ. According to Anaximander, every thing has an opposite. It follows that there are only two opposites that can be produced from nothingness. Tentatively, we can justify the appearance of energy from nothingness by assuming that this energy is of two kinds, kinetic (positive) and potential (negative). Unfortunately this raises other difficulties, and in any case it is no real explanation. The only really satisfactory answer seems to be that the human spirit *gives existence to things*, formulating them as pairs of opposites. In this case, the creation of the universe is brought back to the creation of the mind, the creation of thought, the creation of God – in the broadest sense. Once more there is a reconcilation with Scripture: 'In the beginning was the Word'.

All this shows that with regard to his intuition of these fundamental problems, man has made depressingly little progress. All we can do is to use our intuition as well as we can, basing it upon the state of knowledge at the present time.

2 *The absolute temperature of the universe* As we have seen, the universe must have originally been at a very high temperature. By this we mean several thousands of millions of degrees (it is generally estimated that the value was about a

million million degrees absolute). Matter did not yet exist; the whole of space was filled with radiation at a temperature several thousand million times greater than that in the core of an exploding H bomb. Space was of small dimensions, so that the idea of a kind of super-H bomb is not so far-fetched as might be thought. If theory and observation are to be trusted, the universe began in an apocalyptic state not unlike that of an atomic explosion – apart from the fact that there was no matter. This initial phase is often called the Big Bang. It has also been called the primaeval fireball. The Ancients would have been able, with good reason, to compare it with the gates of the Inferno!

The radius of the universe grew, and the expansion was accompanied by a steady cooling of the thermal radiation which made up the universe. The temperature of space fell rapidly, until it had dropped to the values to which we are more accustomed. Theoretically, this drop in temperature must have been quick. Before the universe was one thousand years old (which, after all, is very little when we remember that its actual age is now over 10,000 million years), the general temperature was about 300° absolute, or about 27° Centigrade. A thousand years after its birth, the universe became habitable. Subsequently the general temperature fell still further, and by now has reached a value of only a few degrees absolute.

All this is convincing enough in theory, but it is only during the past few years that we have had any direct experimental evidence for the Big Bang. Some cosmological theories even maintained that the universe had no beginning at all, but has always existed in the state which we know today; the continuous creation theory is an example of this. Therefore, it is obviously important to see whether we can find any traces of the initial Big Bang, and this was the task of experimental cosmology in the years following 1960. It can be noted that the 'remains' of the original explosion have been found, in the form of a temperature of about 3° absolute for the whole of cosmic space. At the present time, the entire universe is thus bathed in homogeneous, isotropic radiation (black-body radiation) at 3° absolute.

This is an important result. To use an analogy given by the American physicist John Wheeler, we can compare the 10,000 million years since the Big Bang with the height of the Empire State Building (112 floors); an observer on the top of the building, looking downward, will find that the most remote galaxies visible optically are at about the 60th floor and the most distant quasars on the 20th floor, but now, with the new information drawn from the 3° of temperature, our observer can see down to the first floor – or even into the street.

In addition to all this, the value of the black-body temperature of the universe seems to show that the universe is finite. Otherwise, it is hard to see how thermodynamic equilibrium could apply; in an open universe, the radiation could spread outward indefinitely in all directions. If the evidence for a 3° temperature is accepted, space would seem to be spherical, since this is the only type of space which is finite. Moreover, this is a confirmation of what Albert Einstein had intuitively supposed; the universe is like our Earth inasmuch as one could travel right round it and come back to the point of departure. Of course, the Earth's surface can be described in two dimensions, space in three, and space-time in four; but in any event, our new information about the original Big Bang, and its results, are of great importance cosmologically.

The future

It seems, then, that we can be reasonably certain of the past history of the universe; in the beginning it was very small and extremely hot, but during the 10,000 million years that followed its birth it expanded steadily, and the temperature dropped to its present-day value of only 3° absolute. During this time, matter was produced out of the original radiation, and gravitational forces did the rest, building up suns and planets. The most important steps in the evolution of the cosmos were the changing of radiation into matter, matter into life, and life into man. It follows that the human phenomenon is spread over the universe in the same way as the phenomena of life, matter and radiation.

So much for the past; what, then, about the future of the universe? We must try to make up our minds whether it will evolve to a predetermined state, or whether evolution will follow a generally defined direction with random variations.

The question is not easy to formulate if it is to be taken in all its profound significance. We must decide whether the phenomena which we observe at the present moment have been determined solely as a result of past events, or, at least, of the relationships of cause and effect (no matter whether these laws be probabilistic or deterministic), or whether there is a well marked line through time which links the past, present and future.

To help to appreciate this dilemma, consider a skier who is going down the slopes of a mountain. He is at liberty to choose whatever path he likes, and to perform any tricks of which he is capable; he can also stop whenever he chooses. However, he is compelled to keep on descending, following the general slope of the mountainside, so that at the end of his ride he is bound to be at a lower level than the starting point. In this case, the skier's twists and turns as he goes along the ski-trail are of his own choosing; they are produced by the relationships of cause and effect, and the free choice of the skier is all-important. Yet on the scale of the whole trail, the skier is not free to make up his own mind. He cannot climb back up the slope, and he must follow the general form of the mountain, so that he is bound to end his ride in the valley below.

It is reasonable to ask whether the evolution of the universe can be of the same kind, and must follow a course which will lead to an inevitable end which is dependent upon the form of the universe in its entirety – that is, the general form of space-time. If so, the universe had a well defined beginning, and will have an equally well defined end, with all the observed phenomena having a trend in this direction.

Not surprisingly, optical astronomy gives us our best method of studying Relativity in connection with the universe considered as a whole. The cosmological model now most generally favoured involves a pulsation of the radius of the universe. We are now at an epoch when the radius is increasing, so causing the observed expansion of the universe, but this will not continue indefinitely. After reaching a maximum value, the radius will begin to shrink once more, so that the galaxies will approach each other instead of receding; as the distances grow less, the temperature of space will rise. The end of the world will come in a tremendous cataclysm, a general collision of the galaxies meeting each other; the temperature will rise rapidly to a million, then to a thousand million degrees. It is not enough to say that at this moment all life will have vanished from the universe. Matter itself will be reconverted into radiation, and everything will return to light. At last the universe will have nil, or almost nil, volume, so that it will to all intents and purposes have disappeared.

However, our cosmological model does not end its forecasts here. It goes on to indicate that after having passed through this extreme phase, the universe will be ready to start afresh. Expansion will begin, and there will be the gradual reappearance of matter, stars, planets, life and man. This reminds one of the Great Year in which the Ancients believed; they thought that at the end of this Great Year everything would begin again, in a new cycle which was identical with the old. The difference between this idea and our own is no more than quantitative. A century before Christ, Cicero forecast that the Great Year would come in 12,954 years; modern cosmology puts it at between 20 and 25 thousand million years hence, but the principle is the same. It is not easy to decide whether this is a fundamental intuition of the Ancients or, on the contrary, a failure of imagination on the part of the Moderns!

An evolutionary cosmology

It is true to say that there is more to twentieth-century cosmology than the idea of a cyclic universe, and that in the course of time the various situations will simply repeat themselves. First, there is the mental outlook which has become more and more perceptive during the last few decades, and which is therefore the result of a relativistic method of thinking. The universe can no longer be described in an absolute manner. New concepts have been introduced into our language; we have our basic axioms, the mathematical and logical methods which depend upon these axioms, and we have the practical observations. All these must be taken together with the mechanism of human thought.

The cosmological universe at the beginning of the twenty-first century must, therefore, be a universe of the mind. Of course, it is very interesting to know that the universe is pulsating, and that its radius oscillates perpetually between zero and a maximum value, but this result is not absolute; it is no more than a framework inside which the human mind can collect his knowledge together into a coherent plan. In the future, new knowledge and new observations may alter this elaborate plan, but this depends upon the development of mathematical techniques and advanced concepts.

There is another strong reason why the cosmology of the end of the twentieth century must be more elaborate than the simple idea of a pulsating universe. We must regard the effect of time as non-reversible. Quite apart from the behaviour of matter itself, this can be shown by the evolution of a star. During the earlier stages a star changes its hydrogen into helium; then comes the period of old age, when the star has exhausted its hydrogen, and ends its separate existence by a more or less catastrophic process (possibly a supernova explosion). There is no reversibility here. A star's old age is not followed by renewed youth.

But it is with living material that this non-reversibility of time is most striking. In a living cell there is the process of orientation, so that chaos becomes steadily more and more orderly until the molecules which make up the living material are working together toward a final state as though there is a sort of arrow which passes through time and makes a clear distinction between the future and the past. The mechanisms of living matter remain totally inexplicable in the light of the simple physical and chemical laws if we assume that time is reversible – that is, if we do not agree that the universe is evolving; it began 10,000 million years ago, and will end up in a different state in the future. The simple elements of the matter are capable of knowing how, as in living material, they can organise themselves so as to take part in the evolution toward such a state. In short – and to come back to more conventional physical terms – everything seems to show that the great universal laws, which apply everywhere, are restricted by the limiting conditions for the states of space-time which are defined not only in the past, but also in the future.

It will be fitting to end this chapter with some words written by Pierre Teilhard de Chardin, in which the great visionary is putting forward his conviction that there is a Cosmology in which, without doubt, the whole of the universe is evolving toward its destiny:

Life may be compared with a snowball; in its protoplasm, character is heaped upon character. It becomes more and more complicated.... But how is this movement of expansion shown, as a whole? Is it a controlled explosion, like that in a motor, or is the explosion triggered off in some uncontrolled manner?

No researchers nowadays doubt that evolution is going on, but there is no agreement as to whether or not this evolution is directed. Ask a biologist of today whether he admits that life has a connected thread throughout all its transformations; nine times out of ten he will answer 'No', and he may even be vehement about it. You can then point out that the organised material is in a state of continual metamorphosis, and may develop toward the most unlikely forms – as is evident at a glance. Yet how can we decide upon the absolute, or even relative, values of these fragile constructions? By what right do we say that a mammal, or even a man, is more advanced and more perfect than a bee or a rose? The solutions are diverse, and yet equivalent ... because nothing seems to go back to nothingness.

240

Science in its ascent, and humanity in its forward march, seem to mark time at this point, because the mind is slow to recognise that there is a precise orientation and a privileged axis of evolution.... Weakened by this fundamental doubt, the researchers are unable to come to any valid conclusions ... I believe that I can see that in life there is a sense and a line of progress – so well marked, indeed, that I have convinced myself that their reality must be universally admitted by the science of tomorrow.

14 Conclusion

It is not easy to sum up these last twenty-five centuries of cosmology in a kind of balance sheet, because human thought is much too advanced to draw up anything so mundane as a balance sheet when dealing with the fundamental problem of the universe as a whole.

In the preceding pages, we have limited ourselves to only a very few of the cosmological ideas that have been put forward over the ages; it would have been very instructive also to have examined cosmologies of the Orient, of the ancient peoples of America, of the natives of Africa. But in most of these various cosmologies – more specifically those which have been reviewed here – there are certain common aims of research, two of which are of particular importance. These deal with form and with number.

Man has tried to find the form of the universe, considered in its entirety, rather in the manner of an artist considering his model in line and in colour. To find out this form, man had to make the same choice as a sculptor; he had to find the basic substance upon which to work. Therefore he drew up axioms, and from these worked out various essential principles which were assumed to be applicable everywhere.

Very naturally, we turn at this juncture to geometry. From Plato to Einstein, bypassing Kepler and Descartes, we follow up this idea of a design for the universe, so that it can be represented in visual form. The so-called perfect curves, together with the regular solids, played a vital role in the investigation; the universe was assumed to be perfect, since it was God's creation. Einstein went back to, and refined, this idea of a geometrical description of the universe. The cosmological models founded upon Relativity represent the whole of space-time as a geometrical figure in four dimensions.

As well as following this geometrical approach, man has done his best to describe the universe in terms of numbers. From Pythagoras to the Copenhagen school, the concept of number has held a dominant place in the mechanism of human thought in its observation and judgment of the outer world. These numbers can be translated in various ways, but the whole or natural numbers play a fundamental role; many thinkers over the ages have shared the view of the celebrated mathematician Kronecker, who declared, in the nineteenth century, that 'God created the natural numbers and man created the rest'. Numbers are also involved in measurements, which, since the time of Newton, have become the corner-stones upon which the whole of modern science has been built.

If it is thought useful to define a line of progress in cosmological vision over the course of the centuries, it seems that we can sum it up in a single word: Relativity.

The Ancients believed in an absolute universe, created by God, with the Earth lying at rest in the centre of the sphere of the fixed stars. Little by little the Earth was demoted to its proper place among the planets; then the Sun itself, originally thought to lie at the centre of the Milky Way, was trans-ferred to the edge of one of the galactic arms. It was originally believed that our Galaxy, with its 100,000 million stars, was alone in the universe; then it was discovered that galaxies similar to our own exist in their thousands of millions. Planets of Earth-type also exist in thousands of millions, and by now it is scarcely possible to claim scientifically that Earthmen are unique; there must be thousands of millions of human races in the cosmos.

By now, then, it seems that our whole point of view of the universe must be relative. It is not false to claim that the

Earth is stationary, and yet it is turning relative to the Sun; it is not false to claim that the Sun is motionless, and yet it is moving relative to the Milky Way. This applies equally to the galaxies, and is valid for all the phenomena of nature.

It might have been hoped that now, when a complete plan of the universe seems to be within our grasp – thanks to the cosmological models drawn from Relativity – it would have been possible to give an absolute description of the phenomena of the universe, since these phenomena form what may be called a 'rigid block' in the plan of space-time. In this plan, nothing would evolve, since past, present and future would figure simultaneously in it. Everything would exist throughout the whole of time.

But in following Relativity through to this extent, it becomes clear that such a view is not tenable. The plan of the universe which this theory would give us would apply relatively to present Man, and to his current knowledge. Subsequently the whole of the plan will change; past and future will crystallise in a different manner around the knowledge of tomorrow. Yesterday the universe was 4,000 years old; today the age is 12,000 million years; what will it be tomorrow? We must wait and see; but what we do know is that the answer to the question remains relative – relative to man who formulates it.

It is, therefore, no exaggeration to say that by the end of the twentieth century, cosmology will be, above all, spiritual in nature. It is for man that man succeeds in formulating his ideas; his consciousness is evolving. The universe is language; it is thought; it is, finally, the Word. All through the progress of cosmological thought, we have been able to make a better appreciation of the profound significance of the intuition 'In the beginning was the Word'.

Bibliography

Bondi, H. 1960. *Cosmology*, Cambridge University Press, London/ New York.

Bondi, H., Bonnor, W.B., Lyttleton, R.A. and Whitrow, G.J. 1960. *Rival Theories of Cosmology*, Oxford University Press, London/New York.

Bonnor, W. 1964. *The Mystery of the Expanding Universe*, Eyre & Spottiswoode, London/Macmillan, New York.

Butler, S.T. and Messel, H. (eds.) 1966. *Atoms to Andromeda*, Pergamon Press, London/New York.

Cayeux, A. de, 1964. *Trois milliards d'années de vie*, Encyclopédie Planète, Paris.

Charon, J. 1961. *La connaissance de l'Univers*, Seuil. Paris.

Charon, J. 1962. *Du temps, de l'espace et des Hommes*, Seuil, Paris.

Charon, J. 1963. *15 leçons sur la Relativité générale*, Kister, Genève/ Ed. de la Grange Batelière, Paris.

Charon, J. 1968. *Man in search of himself*, Walker, New York.

Charon, J. 1969. *Cours de Théorie relativiste unitaire*, Albin Michel, Paris.

Childe, V.G. 1958. *The Prehistory of European Society*, Penguin Books, London/Baltimore.

Einstein, A. 1934. *Mein Weltbild*, Amsterdam. Translated 1935 by Alan Harris as *The world as I see it*, John Lane, London.

Einstein, A. 1952. *Conceptions scientifiques, morales et sociales*, Flammarion, Paris.

Graham-Smith, F. 1960. *Radio-Astronomy*, Penguin Books, London/Baltimore.

Grousset, R. 1946. *Bilan de l'Histoire*, Plon, Paris.

Hazard, P. 1961. *La crise de la conscience européene de 1680 à 1715*, Fayard, Paris. Translated as *European Mind, 1680–1715*, Peter Smith, Magnolia, U.S.A.

Heisenberg, W. 1958. *Physics and Philosophy, the revolution in modern science*, Allen & Unwin, London/Harper, New York.

Heisenberg, W., Born, M., Schrodinger, E. and Auger, P. 1961. *On Modern Physics* (English translation of 'Sulla fisica moderna' by M. Goodman and J.W. Binno, Orion Press, London).

Koestler, A. 1959. *The Sleepwalkers, a history of man's changing view of the universe*, Hutchinson, London/Macmillan, New York.

Laborit, H. 1963. *Du Soleil à l'Homme*, Masson, Paris.

Leithauser, J.G. 1955. *L'Homme à la conquête de l'Univers*, Plon, Paris.

Lovell, B. 1962. *The Exploration of Outer Space*, Oxford University Press, London/Harper, New York.

McVittie, G.C. 1961. *Fact and Theory in Cosmology*, Eyre and Spottiswoode, London/Macmillan, New York.

Moore, P. 1954 and 1968 (revised). *Suns, Myths and Men*, Frederick Muller, London.

Munitz, M.K. 1957. *Theories of the Universe*, Free Press, Glencoe.

Oppenheimer, J.R. 1954. *Science and Common Understanding*, Simon and Schuster, New York.

Piddington, T.H. 1961. *Radio Astronomy*, Hutchinson, London.

Planck, M. 1963. *L'image du Monde dans la Physique moderne*, Gonthier, Paris.

Prat, H. 1960. *Metamorphose explosive de l'Humanité*, Sedes, Paris.

Schatzman, E. 1965. *The Origin and Evolution of the Universe*, Basic Books, New York/Hutchinson, London.

Schatzman, E. 1968. *The Structure of the Universe*, Weidenfeld and Nicolson (World University Library), London/McGraw-Hill, New York.

Sciama, D.W. 1959. *The Unity of the Universe*, Faber, London/Doubleday, New York.

Singh, J. 1961. *Great Ideas and Theories of Modern Cosmology*, Constable, London/Dover, New York.

Teilhard de Chardin, P. 1959. *The Phenomenon of Man*, Fontana Books, London/Harper, New York.

Whitrow, G.J. 1955. *Structure and Evolution of the Universe*, Hillary House, New York.

Whittaker, E. 1942. *The Beginning and End of the World*, Oxford University Press, London/New York.

Zafiropulo, J. 1961. *Apollon et Dionysos*, Les Belles Lettres, Paris.

Acknowledgments

Acknowledgment is due to the following for the illustrations (the numbers refer to the page on which the illustration appears).

40, 41, 62, 63 Magnum Photos, photo Eric Lessing; 42, 47, 89, 93, 107, 124, 147, 162, 170, 189, 198, 221 courtesy Patrick Moore; 46 Planetarium Magazine; 61, 79, 123 Mansell Collection; 88, 141 Science Museum, London; 105–6 H. Hatfield; 143 The National Trust, photo David Swann; 158–9, 161, 218 Royal Astronomical Society; 160 'Crown copyright' Science Museum, London; 166 Yerkes Laboratory; 172–3, 177 Mount Wilson and Palomar Laboratories; 183 Lotte Jacobi; 212–3 NASA; 214–5 Lund Observatory.

The diagrams were drawn by Design Practitioners Limited.

250

254

scientists 15, 35, 78, 142, 225, 230
Scripture 108, 109, 110, 121, 122, 125, 157, 230, 232
semantics 133, 135
Sfondrati, Cardinal 108
Schönberg, Nicolas, Cardinal of Capua 68
Sidereal Messenger, The, see Galileo
Simocatta, Letters of 66
Simplicio 114, 115, 116, 121
Sirius 157, 171
Sitter, de 196
Slipher, V. M. 171, 175
Solar System 86, 164, 209, 210, 211, 220
space 26, 29, 35, 122, 133, 134, 135, 136, 149, 150, 151, 181, 182, 222, 233, 234; and time 10, 185, 203, 206, 207, 225, 229; curvature of 188, 190, 195, 196; absolute 221, 225–7 space-time 131, 132, 133, 134, 136, 151, 182, 188, 190–1, 222, 228, 234, 236, 239, 241
space travel 209 *passim*; and acceleration 219–27; and problem of ageing 225–7
Spain 55
spectroscopy 167, 171; *and see* galaxies, red shift in spectrum
spheres, and Aristotle 32, 39, 51, 59; sphere of fixed stars 22, 26, 73, 242, spherical universe 21, 28–30; and Kepler's cosmology 82, 84
stars 11, 52, 54, 122, 157, 190, 196, 197, 206; and Aristotelian cosmology 32, 39; and Aristarchus 35; movements of, and Herschel 157, 163–4; distances 167, 171–9;

and space travel 210 *passim; see also* binary stars, Cygni, Cepheids, Hubble, Proxima Centauri, Sirius
Stendhal, *Pensées diverses* 57
sublunar region, of Aristotle 32, 122
Sun 22, 26, 30, 57, 86, 99, 104, 179, 230, 242, 243; in Aristotelian system 32, 39; and heliocentric universe of Aristarchus 34–6, 65; and Ptolemaic system 44, 48, 50; and eclipses 45; and Saint Augustine 53; and Copernican system 60, 65, 68, 73–4, 78, 104, 114; and Galileo's theory of tides 109–10; and Kepler's cosmology 82–92; sun spot 99; and Cartesian cosmology 134–5; and gravity 157, 163; in Herschel's star system 164, 167; spectrum of 167; and Relativity 188–98
supernova 122
super-observer, of universe 14–5

Teilhard de Chardin, Pierre 15, 239–40; *Le Phénomène Humain* 15
Tennyson, Alfred Lord, 'Enoch Arden' 13
Thales of Miletus 31
theology 80, 103
theorems 130
theory, and observation 27, 51, 77
thought, human 11, 119, 127, 152, 184, 228, 232, 238, 241, 242
tides, Galileo's theory of 109, 114–5
time 181–2; *see also* space, space-time
trigonometry 86
Tübingen, University of 80

The Quest for Absolute Zero

K. Mendelssohn

According to the laws of physics the absolute zero of temperature can never be reached, but it can, in principle, be approached to an arbitrary degree. Louis Cailletet's liquefaction of oxygen in 1877 started a quest for ever lower temperatures which was pursued in laboratories all over the world. More recently absolute zero has been approached to within a few millimetres of a degree centigrade. The phenomena observed at these very low temperatures have led not only to new concepts in physics but also to the beginning of a new technology. The absorbing story of the quest for absolute zero is told here for the first time by an author who through four decades has himself been in the forefront of low temperature research.

Some international comments

'Unusual and utterly fascinating . . . it is intended for undergraduate students of physics, but should be on the bookshelves of a much larger group of readers'   *American Scientist* **(USA)**

'La historia apasionante de la busca del cero absoluto se cuenta por primera vez en este libro por un autor que durante cuatro décadas ha ocupado un lugar sobresaliente en la investigación de las bajas temperaturas y para quein los acontecimientos relatados han constituido una experiencia personal'   *La Nación* **(Argentina)**

'Voor ieder die geinteresseerd is in fysica, is dit een fascinerend boek'
Arnhemse Courant **(Holland)**

'Det är en synnerligen välskriven bok'   *Biblioteksbladet* **(Sweden)**

'Nell'insieme un libro di divulgazione scientifica indubbiamente positivo'
Sapere **(Italy)**

'A remarkable book, as absorbing as a novel and as instructive as the best text-book . . . a classic'   *Chemical Engineer* **(UK)**

#3